AF560466

# Nutritional and Ecological Requirements of Forest

# Nutritional and Ecological Requirements of Forest

Ajay Mukherjee

RANDOM PUBLICATIONS
NEW DELHI (INDIA)

**Nutritional and Ecological Requirements of Forest**

ISBN 978-93-5111-610-3



Published in 2015 in India by

**RANDOM PUBLICATIONS**

4376-A/4B, Gali Murari Lal, Ansari Road
New Delhi-110 002
Phone : +9111-43580356, 011-23289044, 011-43142548
e-mail: sales@randompublications.com,
info@randompublications.com, randomexports@gmail.com

Reprinted 2019

*Type Setting by* : Friends Media, Delhi-110089
*Digitally Printed at* : Replika Press Pvt. Ltd.

# Preface

A Forest is a large area of land covered with trees or other woody vegetation. Hundreds of more precise definitions of forest are used throughout the world, incorporating factors such as tree density, tree height, land use, legal standing and ecological function. According to the widely-used United Nations Food and Agriculture Organization definition, forests covered an area of four billion hectares or approximately 30 percent of the world's land area in 2006.

Forests are the dominant terrestrial ecosystem on Earth, and are distributed across the globe. Forests account for 75% of the gross primary productivity of the Earth's biosphere, and contain 80% of the Earth's plant biomass. Forests at different latitudes form distinctly different ecozones: boreal forests near the poles tend to consist of evergreens, while tropical forests near the equator tend to be distinct from the temperate forests at mid-latitude. The amount of precipitation and the elevation of the forest also affects forest composition. Forests account for 75% of the gross primary productivity of the Earth's biosphere, and contain 80% of the Earth's plant biomass.

Forest ecosystems can be found in all regions capable of sustaining tree growth, at altitudes up to the tree line, except where natural fire frequency or other disturbance is too high, or where the environment has been altered by human activity. The latitudes 10° north and south of the equator are mostly covered in tropical rainforest, and the latitudes between 53°N and 67°N have boreal forest.

As a general rule, forests dominated by angiosperms are more species-rich than those dominated by gymnosperms, although exceptions exist. Forests sometimes contain many tree species only within a small area or relatively few species over large areas. Forests are often home to many animal and plant species, and biomass per unit area is high compared to other vegetation communities. Much of this biomass occurs below ground in the root systems and as partially decomposed plant detritus. The woody component of a forest contains lignin, which is relatively slow to decompose compared with other organic materials such as cellulose or carbohydrate.

I would like to thank my team for standing beside me throughout my career and writing this book. My special thanks go to "Random Publications" who have published the book.

*– Ajay Mukherjee*

# Contents

# 1

# Understanding and Managing Conservation Conflicts

The crisis in tropical forest resources is today recognised as the world's most severe ecological and economic crisis. This is primarily because of the linkages between the genetic resources and diversity of tropical forests and food security of the entire world, and the linkages between the ecological stability of tropical forests and the economic well-being of the majority of the world's people who live in the tropics, in what is called the Third World.

These Third World countries, the erstwhile colonies of the industrialized coun tries, provided natural resources on which the industrialization of the latter was based. Both industrialisation and economic growth in the colonial and post-colonial periods have been based on the reckless exploitation of tropical forests. Today, the cumulative impact of this over-exploitation has led to critical and almost irreversible ecological degradation. The famine in Africa and other arid regions has shifted the world's attention to the high ecological and social costs of tropical deforestation. It has become the central concern of governments, development agencies and ecology movements. Yet the focus on tropical deforestation and its reversal is not automatically translated into the protection of tropical forests and those who depend on them for survival. As long as misconceptions about the nature of tropical forest ecosystems prevail, as long as conflicting demands are ignored and as long as the causes of tropical deforestation are inaccurately located, this degradation cannot be arrested.

As a result, tropical ecosystems will continue to be degraded, the survival of people of the Third World will continue to be threatened, and forest conflicts will continue to grow. Normally, forests are identified only with the economy associated with the commercial-industrial use of forests, however, the crisis of tropical forestry needs to be understood in the light of the conflicting demands for forest biomass, by the three fundamental economies associated with forests:

1. Nature's economy of essential ecological processes.
2. The survival economy of basic needs satisfaction of the people.
3. The market economy of industrial-commercial demands.

Nature's economy of essential ecological processes generates a demand on the products of the forest in terms of the maintenance of the stability of soil systems and the hydrological balance of the forest ecosystems. In ecologically sensitive ecosystems like upland watersheds, this economy becomes the most crucial one and should be given the necessary attention in forest management. Neglecting this economy in upland watersheds will imply tremendous negative externalities to the national exchequer as relief for regular floods and drought, which are easily described as loss due to nature's fury.

The survival economy of basic needs satisfaction reflects the requirement of forest biomass of the people living in and in the vicinity of forests in terms of fuelwood, fodder, fruits, nuts, green manure, small timber, etc. In forest areas where human settlements have either existed or are in the vicinity, the requirements of the survival economy have been satisfied all along without any major ecological damage. However, under certain situations, the pressure of the survival economy can be substantial and its neglect can lead to the unexpected and rapid degradation of forest resources.'

The market economy of industrial-commercial demands coastitutes the forest biomass demand of the total market system in the formal market economy. It includes the demand for pulpwood, plywood, furniture, house construction, etc., as well as fuelwood for the urban market. The demand for biomass from the urban industrial sector as well as the survival requirements of people have increased dramatically in the last century in India. There are many examples which illustrate that the growth of forest based industries is disproportionately beyond the ecological limit of the renewable productivity of nature under the present system of management. The growth in population adds another significant demand for biomass for domestic purposes.

It has, however, not been recognised, quantified and internalized in formal forest management. In the perspective of the increasing demand from survival and market economies, the biomass requirement for nature's economy are systematically sacrificed and nature's needs remain totally unfulfilled. In due course, this leads to the ecological destabilisation of the forest ecosystems.

Again, the conflict between survival and market economies assumes large proportions when most of the biomass produced is cornered by the economically powerful groups through the formal official mechanisms while the basic needs of fuel, fodder and small timber of the economically weak remain unsatisfied due to their weak political and economic status. Obviously, these three diverse economies push for the satisfaction of their demands either silently or loudly, resulting in both overt and covert conflicts over forest resources.

The tacit and invisible nature of these conflicts combined with a mistaken or incomplete understanding of these diverse economies can result in the lack of perception of conflicts between them. The failure to perceive conflicts in forest use can only aggravate the forestry crisis. It is thus absolutely essential to understand and perceive the covert as well as explicit demands on forest resources to evolve a forest policy that is ecologically sensitive and socially just, so that our forest and land resources can be used for the overall satisfaction of the needs of the nation and the people in an equitable and sustainable manner.

The conceptual framework based on the three economies adopted here differs from the conventional categories related to forest resources. We do not use the dichotomy between quantity/quality as categories of forest use that Hallsworth, for example, has used. According to him, Human demands on the forest fall into two categories. First a demand for forest quantities; for the actual things a forest can produce: for timber, for food and for space for cultivation or for grazing; secondly, a demand for forest qualities; for the effects that forests have on the environment of man-to protect supplies of water, provide havens for wild-life and maintain the pool of genetic resources, to protect the soil against erosion, and to provide space for recreations.

The dichotomy between quantity and quality, as between tangible and intangible, fails to capture the reality of those village women in Garhwal who launched the Chipko movement because for them water from the forest is a more significant product than wood and timber. For them water is not an 'intangible' produce or a mere quality. It is more tangible and basic in their sustenance economy than the commercial wood extracted and exported from their forests.

For similar reasons we have not adopted the local-national dichotomy between local and national interests to understand conflicts over natural resources, because the local subsumes the national interest. The national interest is or should be the integration of all local interests, not just another interest acting over and against local interests.

The demands of nature's economy and survival economy are national demands, seen throughout the country, and are not peculiar to a particular region. What has usually passed as the 'national interest' are the large scale demands of the commercial industrial sector, operating through the government. The emergence of ecology movements at the national level are indicative of the fact that these are demands of a socially narrow sector which come into conflict with socially broader sets of demands that are more appropriately termed the national interest.

We have also avoided the categories of private versus public demands because they fail to show adequately that what are called individual or private needs at the local level are located in the public context of a village community, and what usually passes as public policy in forestry is a government policy which is totally subservient to commercial/industrial interests. The 'privatisation' inherent in a commercially-oriented forest policy is What has been challenged by movements like Chipko which demand a more effective social and public control over forest resources. The official management of forest resources in the name of 'national' or 'public' interest has, in reality, privatised people's 'common resources. This privatization trend which was rooted in the colonial period has survived to the present day. As Chhattarpati Singh observes, 'The class or "public" which more often than not benefits from such acquisitions is the rich. This is patently true in the acquisition of common land, especially forests. The public of the "public purpose" for which forests have been acquired constitutes all but the forest dwellers.

We have therefore preferred to distinguish human demands on the basis of use for survival and use for commerce and industry, and have included nature's demands to avoid an anthropocentric bias as well as to have a context to locate forest movements like Chipko which struggle for forest conservation to respect nature's rights, and not just to assert their own rights to forest resources.

## ECOLOGY MOVEMENTS AND CONFLICTS OVER NATURAL RESOURCES

The recent period in human history contrasts with all the earlier ones in its strikingly high rate of resource utilization. Ever expanding and intensifying industrial and agricultural production has generated increasing demands on the world's total stock and flow of resources.

These demands are mostly generated from the industrially advanced countries of the North and the industrial enclaves in the underdeveloped countries of the South. Paradoxically, the increasing dependence of the industrialised societies on natural resources, through the rapid spread of energy and resource-intensive production technologies, has been accompanied by the spread of the myth that increased dependence on modern technologies implies a decreased dependence on nature and natural resources This myth is supported by the introduction of a long and indirect chain of resource utilisation which leaves invisible the real material resource demands of the industrial processes.

Through this combination of resource intensity at the material level and resource indifference at the conceptual and political levels, conflicts over natural resources generated by the new pattern of resource utilisation are generally shrouded and overlooked. These conflicts become visible when resource and energy-intensive industrial technologies are challenged by communities whose survival depends on the conservation of resources threatened by destruction and overexploitation, or when the devastatingly destructive potential of some industrial technologies is demonstrated as in the Bhopal disaster.

For centuries, vital natural resources like land, water and forests had been controlled and used collectively by village communities thus ensuring a sustainable use of these renewable resources. The first radical change in resource control and the emergence of major conflicts over natural resources induced by non-local factors was associated with colonial domination of this part of the world. Colonial domination systematically transformed the common vital resources into commodities for generating profits and growth of revenues. The first industrial revolution was to a large extent supported by this transformation of commons into commodities which permitted European industries access to the resources of South Asia.

With the collapse of the international colonial structure and the establishment of sovereign countries in the region, this international conflict over natural resources was expected to be reduced and replaced by resource policies guided by comprehensive national interests. However, resource use policies continued along the colonial pattern and, in the recent past, a second drastic change in resource use has been initiated to meet the international requirements and the demands of the elites in the Third World, leading to yet another acute conflict among the diverse interests.

The most seriously threatened interest, in this conflict, appears to be that of the politically weak and socially disorganised group whose resource requirements are minimal and whose survival is primarily dependent directly on the products of nature outside the market system. Recent changes in resource utilisation have almost wholly by-passed the survival needs of these groups. These changes are primarily guided by the requirements of the countries of the North and of the elites of the South.

This book analyses environmental conflicts in contemporary human society. In general it relates to societies all over the world, but in particular it addresses the most intense and emerging social contradictions in India related to conflicts over natural resources. Science and technology are central to these conflicts because while scientific knowledge has been used by contemporary societies to considerably enlarge man's access to natural resources, it has also allowed the utilisation natural resources at extremely high rates.

The contemporary period is characterised by the emergence of ecology movements in all parts of the world which are attempting to redesign the pattern and extent of natural resource utilisation to ensure social equality and ecological sustainability. Ecology movements emerging from conflicts over natural resources and the people's right to survival are spreading in regions like the Indian subcontinent where most natural resources are already being utilised to fulfil the basic survival needs of a large majority of people.

The introduction of resource and energy-intensive production technologies under such conditions leads to economic growth for a small minority while, at the same time, undermines the material basis for the survival of the large majority. In this way, ecology movements have questioned the validity of the dominant concepts and indicators of economic development. The ideology of economic development, which remained almost monolithic in the post World War II period, is thus faced with a major foundational challenge.

In this chapter an attempt has been made to provide a systematic conceptual framework for analysing the processes and structures of modern economic development from an ecological -perspective. It attempts to analyse the relationship between economic development and conflicts over natural resources to trace the roots of ecological movements.

Further, in the light of the ecological perspective, it examines the fundamental assumptions and categories of modern development economics that are used to determine the objectives of economic development as well as the criteria for the choice of technologies that are used to achieve these objectives.

## ECONOMIC DEVELOPMENT AND ENVIRONMENTAL CONFLICTS IN INDIA

A characteristic of Indian civilization has been its sensitivity to natural ecosystems, vital renewable natural resources like vegetation, soil and water were managed and utilised according to well defined social norms that respected the known ecological processes. The indigenous modes of natural resources utilisation were sensitive to the limits to which these resources could be used It is said that the codes of visiting important pilgrim centres Badrinath in the sensitive Himalayan ecosystem, included a maximum stay of one night so that the temple area would not put excess pressure on the local natural resources base.

In the pre-colonial indigenous economic processes, the levels of utilisation of natural resources were not significant enough to result in drastic environmental problems. There were useful social norms for environmentally safe resource utilisation and people protested against the destructive use of resources even by kings. A major change in the utilisation of natural resources of India was introduced by the British who linked the resources of this country with the direct and large non-local demands of Western Europe.

Natural resource utilisation by the East India Company, and later by the colonial rulers, replaced the indigenous organizations for the utilisation of natural resources, like water, forest and minerals, that were mainly managed as commons. With the establishment of British colonial rule in India, the ever increasing resource demands of-the industrial revolution in England were largely met from colonies like India.

Forced cultivation of indigo in Bengal and Bihar, cultivation of cotton in Gujarat and the Deccan led to large-scale commitment of land for the supply of raw materials for the British textile industry, the flagbearer of the industrial revolution. Forests in the sensitive mountain ecosystems like the Western Ghats or the Himalayas were felled to build battleships, or to meet the requirements of the expanding railway network. Forests of the Bengal-Bihar-Orissa region were used for running wood fuel locomotives in the early stages of railway expansion. The latter stages of colonial resource utilisation and control included the monopolization of water rights as in the Sambhar Lake of Rajasthan or the Damoda' Canal in Bengal. Colonial intervention in natural resource management in India led to conflicts over vital renewable natural resources like water or forests and induced new forms of poverty and deprivation.

Changes in resource endowments and entitlements introduced by the British came into conflict with the local people's age old rights and practices related to natural resource utilisation As a result local responses were generated through which people tried to regain and retain control over local natural resources. The indigo Movement in Eastern India, the Deccan Movement for land rights or the forest movement in all forest areas of the country, the Western Ghats, the Central Indian Hills or the Himalayas, were obvious expressions of protest generated by these newly created conflict's.

Conflicts generated by the colonial modes of natural resource exploitation could not, however, grow with a local identity. With the progress of the anti-colonial people's movement at the national level, these local protests merged with the national struggle for independence. With the collapse of colonial rule internationally, and the emergence of sovereign independent countries in the Third World like India, resolution of these conflicts at the local level became a possibility.

While political independence vested the control over natural resources with the Indian state, the colonial institutional framework for natural resource management did not change in essence. Where colonialism collapsed, the slogan of economic development stepped in. There was unfortunately no alternative institutional mechanism other than that of the classical model of development left by the British, with which the newly formed Indian state could respond to the accentuated aspirations of the Indian people for a better life.

The same institutions and concepts, nurtured and developed by the colonial rulers were applied to objectives which were exactly opposite to those of the colonial period. Concepts and categories relating to economic development and natural resource utilisation that had emerged in the specific context of capitalist growth and industrialization in the centres of colonial power were raised to the level of universal assumptions and applicability. The processes which led to deprivation were now entrusted with the responsibility of basic needs satisfaction.

No serious thought was given to the fact that the historical specificity of early industrial development in Western Europe necessitated the permanent occupation of the colonies and the undermining of the local 'natural economy'.' This inexorable logic of resource exploitation, exhaustion and alienation integral to the classical model of economic development based on resource intensive technologies led Gandhi to seek an alternate path of development for India when he wrote:

God forbid that India should ever take to industrialism after the manner of the West. The economic imperialism of a single tiny island kingdom (England) is today keeping the world in chains. If an entire nation of 300 million took to similar economic exploitation, it would strip the world bare like locusts.

While Gandhi's critique was a forewarning against the problems likely to arise by following the classical path of resource-intensive development, at the time of India's independence, there was no clear and comprehensive work plan to realise the Gandhian dream of alternate development that would be resource prudent and would satisfy basic needs.

The issues of resource constraints of economic development were, therefore, not highlighted at the theoretical level, partly due to the tremendous pressure of the enhanced developmental aspirations of a newly independent nation, and partly due to the lack of internalization of natural resource parameters within the framework of economics.

As the scale of economic development activities escalated from one Five Year Plan to another, the disruption of ecological processes that maintain the productivity of the natural resource base started becoming increasingly apparent.

The classical model of economic development in the case of the newly independent nations resulted in the growth of urban-industrial enclaves where commodity production was concentrated, as well as rapid exhaustion of the internal colonies whose resources supported the enhanced demands of these enclaves. In the absence of ecologically enlightened resource management methods, the pressure of poverty enhanced the pace of economic development activities in the hope of a quick improvement in the standard of living for all, as in the case of Western Europe.

For example, commercial forestry earned more revenue by making increasing amount of timber and pulpwood available in the market but in the process reduced the multipurpose biomass productivity or damaged the hydrology of the forests. People dependent on non-timber biomass outputs of forests like leaves, twigs, fruits, nuts, medicines and oils were unable to sustain themselves, in the face of the commercial exploitation of forests. The changed hydrological character of the forests affected both the micro-climate and the stream flows, disturbing the hydrological stability and affecting agricultural production.

There are similar examples from all parts of the country, related to almost all massive developmental interventions in India's natural resource system. Ecological degradation and economic deprivation generated by the resource insensitivity and intensity of the classical model of development have resulted in environmental conflicts, an understanding of which is imperative for the reorientation of our current development priorities and concepts. It is becoming in creasingly clear that these classical concepts and priorities are being used as an alibi to direct 'development' at the national level, while the educated minority elite is the main beneficiary of these 'development' processes.

The ecology movements that have emerged as major social movements in many parts of India are making visible many invisible externalities and pressing for their internalisation in the economic evaluation of the elite-oriented development process. In the context of a limited resource base and unlimited development aspirations, ecology movements have initiated a new political struggle for safeguarding the interests and survival of the poor, the marginalised, including women, tribals and poor peasants.

## ECOLOGY MOVEMENTS AND SURVIVAL

The intensity and range of ecology movements in independent India have continuously widened as predatory exploitation of natural resources to feed the process of development has increased in extent and intensity.

This process has been characterised by the massive expansion of energy and resource-intensive industrial activity and major development projects like large dams, forest exploitation, mining and energy-intensive agriculture. The resource demand of development has led to the narrowing of the natural resource base for the survival of the economically poor and powerless, either by direct transfer of resources away from basic needs or by destruction of the essential ecological process that ensure renewability of the life-supporting natural resources.

In the light of this background, ecology movements emerged as the people's response to this new threat to their survival and as a demand for the ecological conservation of vital life-support systems. The most significant life-support systems in addition to clean air are the common property resources of water, forests and land on which the majority of the poor people of India depend for survival. It is the threat to these resources that has been the focus of ecology movements in the last few decades. Among the various ecology movements in India, the Chipko movement (embrace the trees to oppose fellings) is the most well known.

It began as a movement of the hill people in the state of Uttar Pradesh to save the forest resources from exploitation by contractors from outside.' It later evolved into an ecological movement that was aimed at the maintenance of the ecological stability of the major upland watersheds in India. Spontaneous people's response to save vital forest resources was seen in Jharkhand area in Bihar-Orissa border region as well as in Bastar area of Madhya Pradesh where there were attempts to convert the mixed natural forests into plantations of commercial tree species, to the complete detriment of the tribal people.

In the southern part of India the Appiko movement, which was inspired by the success of the Chipko movement in the Himalayas, is actively involved in stopping illegal over-felling of forests and in replanting forest lands with multipurpose broad leaved tree species. In Himachal Pradesh the Chipko activists have intensified their opposition to the expansion of monoculture plantation of the commercial Chir Pine (Pinus roxburghii). In the Aravalli Hills of Rajasthan there has been a massive programme of tree planting to give employment to those hands which were hitherto engaged in felling of trees.

The exploitation of mineral resources, in particular the opencast mining in the sensitive watersheds of the Himalayas, the Western Ghats and Central India have also resulted in a great deal of environmental damage. As a consequence, environmental movements have come up in these regions to oppose the reckless mining operations.

Most successful among them is the movement against limestone quarrying in the Doon Valley. Here, volunteers of the Chipko movement have led thousands of villagers, in peaceful resistance, to oppose the reckless functioning of limestone quarries that is seen by the people as a direct threat to their economic and physical survival.'

While the Doon Valley instance has a long history of popular opposition to the quarrying of limestone and a Supreme Court order has restricted the area of quarrying to a minimum, examples of such success' of ecology movements are rare People's ecology movements against mineral exploitation in the neighbouring areas of Almora and Pithoragarh still seem to be ignored, probably due to the relative isolation of these interior areas. Beyond the Himalayas, the ecology movement in the Gandhamardan Hills in Orissa against the ecological havoc of bauxite mining has gained momentum and it draws inspiration from the Chipko movement.

The mining project of the Bharat Aluminium Company (BALCO) in the Gandhamardan Hills is being opposed by local youth organisations and tribal people whose survival is directly under threat. The peaceful demonstrators have claimed that the project could be only continued 'over our dead bodies. The situation is more or less the same in large parts of Orissa-Madhya Pradesh region where rich mineral and coal deposits are being opened up for exploitation and thousands of people in these interior areas are being pushed to deprivation and destitution. This is also true of the coal mining areas around the energy capital of the country in Singrauli. In these interior areas of Central India, movements against both mining and forestry are becoming increasingly volatile and people's resistance is growing.

Large river valley projects, which are coming up in India at a very rapid pace, is another group of development projects against which people have organised ecology movements. The large-scale submersion of forest and agricultural lands, a prerequisite for the large river valley projects, always takes a heavy toll of dense forests and the best food growing lands. These have usually been the material basis for the survival of a large number of people in India, specially tribal people. The Silent Valley project in Kerala was opposed by the ecology movement on the ground of its being a threat, not to the survival of the people directly, but to the gene pool of the Tropical Rainforests threatened by submersion. The ecological movement against the Tehri high dam in the UP Himalaya exposes the possible threat to people living both above and below the dam site through large-scale destabilization of land by seepage and strong seismic movements that could be induced by impoundment.

The Tehri Dam Opposition Committee has appealed to the Supreme Court against the proposed dam by identifying it as a threat to the survival of all people living near the river Ganga up to West Bengal. Most notable among the people's movements against dams on the issue of direct threat to survival from submersion are Bedthi lcchampalli, Bhopalpatnam, Narmada Sagar, Koel-Karo, Bodhghat, etc.

In the context of the already overutilised land resources, the proper rehabilitation on a land-to-land basis of millions of people displaced through the construction of dams seems impossible. The cash compensation given instead is inadequate in all respects for providing an alternate livelihood for the majority of the displaced. Destitution is thus the first and foremost precondition for initiating large dam projects

While the process of construction of dams itself invites opposition from ecology movements, the functioning of water projects dependent on the constructed dams results in further ecological disasters and movements. People's movements against widespread water-logging, salinisation and the resulting desertification in the command areas of many dams have been registered.

Among them are instances of protests against the Tawa, Kosi, Gandak, Tungabhadra, Malaprabha, Ghatprabha projects and the canal irrigated areas of Punjab and Haryana. While excess water led to ecological destruction in these cases, improper and unsustainable use of water in the arid and semi-arid regions generated ecology movements in a different way. The anti-drought and desertification movement is gaining momentum in the dry areas of Maharashtra, Karnataka, Rajasthan, Orissa, etc. Ecological water use for survival is being advocated by water based movements like Pani Chetana, Pani Panchayat, and Mukti Sangharsh.

Another major movement originating from the ecological destruction of resources by growth based development is spreading all along the 7,000 km long coastline of India. It is the movement of the small fishing communities against the ecological destruction caused by mechanised fishing whose instant profit motive is destroying the coastal ecology and its long-term biological productivity in a big way. No amount of threat to survival in India from environmental hazards can be complete without a reference to the Bhopal tragedy on 2 December 1984, in which several thousand people died and several lakhs faced serious health hazards following the leakage of poisonous Methyl Iso Cyanate from a pesticide plant of Union Carbide (India) Limited. People's movements for clean air and water are growing in ail parts of the country just as ecologically irresponsible industrialization is moving deeper into the hinterland in search of new resources.

# 2

# Silvicultural and Ecological Considerations

In essence, silviculture is applied forest ecology and forest genetics; it may be described as the development and use of cutting methods and cultural treatments of the forest:

- To ensure adequate regeneration of desirable species as soon as possible after the mature stand is cut,
- To bring about conditions favorable to the optimum yield and quality of production in keeping with the objectives of management and the condition of the forest
- To maintain and where possible improve the quality or productivity of the site.

## REPRODUCTION

A tree crop may be established by artificial or natural means, or by a combination of the two. In artificial reproduction, seed is sometimes sown, but more commonly young trees that have been grown in nurseries are planted on the site. In either case, the seed used is collected in the forest, preferably from trees of good form and vigour. When the crop is established by natural means, the young trees may originate vegetatively or from seed disseminated either before felling takes place or by trees left standing for the purpose at the time of logging. A new crop of young trees that becomes established in the forest before cutting takes place is known as "advance growth."

There are two principal methods of cutting the mature forest, each of which has advantages with respect to the regeneration and establishment of individual tree species:

### CLEAR-CUTTING

By this method, which results in the development of more or less even-aged stands, the forest is cut in one operation or in a succession of operations, each of which removes either a portion of the trees scattered over the whole area or all of the trees from a part of the area; these operations are continued until the original stand is completely cleared. Regeneration may be established by natural or artificial means, the former being provided for either by advance growth, or by seed lying in the surface soil at the time of cutting or made available from the crowns of felled trees or from standing trees surrounding the cutover area.

For species that require overhead protection in their youth a "shelterwood cutting" is made: the stand is opened up gradually over a period of years by making three or four successive cuts periodically over the whole area until the original stand is removed. For other species, the seed source may be the critical factor and the stand is clear-cut except for a few mature trees, evenly distributed over the area, which are preserved to provide the seed for the next crop. This is known as "seed-tree cutting." Quite commonly, portions of a stand are clear-cut periodically in the form of narrow strips, wedge-shaped blocks, or patches, until the stand has been completely felled.

The Douglas-fir forests on the Pacific coast of North America, for example, are clear-cut in patches or blocks of a size that will permit natural regeneration by seeding-in from the surrounding uncut stand.

This type of cutting is particularly applicable to species like the Douglas fir that are intolerant of shade. The intensity of cutting and the size, form and arrangement of clearcut areas vary according to local conditions and requirements, particularly stand density, topography, logging methods and the ecological characteristics of the desired species.

### Selection Cutting

This method distributes cutting and regeneration continuously over the whole area, producing a single uneven-aged stand or series of such stands where trees of all ages from seedlings to mature timber occupy the area together. Scattered single trees or small groups of trees may be removed periodically throughout the whole forest, or the forest may be divided into a number of blocks, one of which is cut over in this way annually or every few years.

The aim is to improve and regenerate the forest, as well as to harvest a crop: while some of the better formed mature trees may be kept in the stand for a number of years to provide seed, all dying and defective trees that are at all merchantable are removed as quickly as possible. Non-merchantable trees may be girdled. As regeneration and early development of growing stock must take place in partial shade, this method is particularly suited to tolerant species.

## AFFORESTATION

Forestry not only seeks to ensure the reproduction of existing forests, but is also concerned with the problem of establishing new forests. Afforestation may be undertaken *(a)* to reclaim forest land cleared for other purposes but best suited for the growing of forest crops, *(b)* to establish forests as an economic crop on land not previously forested, *(c)* to stabilize shifting sand and eroding soils, *(d)* to provide protection for agricultural and other crops against wind and snow damage and *(e)* to improve seepage in drainage areas and so stabilize stream flow.

Great Britain, Australia, New Zealand and South Africa have all undertaken large-scale afforestation with the object of supplementing their natural resources. The trees used have been predominantly softwoods, particularly species of pine. One of the best-known examples of land reclamation for forestry purposes is in the Landes of Gascony, where, since the latter part of the eighteenth century, a programme has been developed to fix shifting sand dunes, to drain swamps and to afforest large areas with maritime pine the principal source of resin and turpentine in Europe.

The use of trees in shelter belts and windbreaks to protect agricultural land is common practice in many countries and has been practiced on a large scale in the United States and Russia. The establishment of forests to conserve water supplies has often been an important objective of forestry ever since its early beginning in Europe. In many cases, however, such forests may be used for wood production as well as for protection. In India and certain other tropical countries forests created primarily to provide small timber and fuel for the local population often help to solve the problems of flood control, erosion and desiccation.

The success of afforestation programmes depends upon many factors. Foresters must determine the suitability of species to soil and climatic conditions; they must guard against the many hazards that may arise in introducing a species or race to a new environment; they must decide whether the species grows best in a pure stand or in mixture with other species and whether it will thrive as a pioneer under open-grown conditions or is better suited to a later stage in plant succession. The relationship of the species to its environment must be studied carefully if afforestation is to be successful.

## "TENDING" OPERATIONS FOR YIELD AND QUALITY

Once the forest is established, the aim of the silviculturist will be to develop and maintain the optimum growth and quality of the stand in keeping with the objectives of management. This is accomplished mainly by cutting operations of various kinds and intensities, especially by means of thinnings carried out periodically throughout the life of the stand. These are commonly referred to as "tending" operations, as distinguished from those aimed primarily at regeneration.

Although forest trees are members of a community and, as such, contribute something towards each other's well-being — *e.g.*, mutual protection against the elements and enrichment of their common habitat, the forest soil — there is, as noted above, a continuous and sharp competition among them for light and space and in some cases for soil moisture and mineral nutrients. The stand density (number of stems per acre) greatly affects the intensity of this competition and also influences the rate of growth per tree and the ultimate value of the forest to man.

Crowded conditions and keen competition are usually beneficial in the very early life of the stand: the inherently vigorous individuals become dominant, the lower branches of the trees remain small and are killed back at an early age and the tree tends to develop a straight bole.Later, however, too much crowding is a disadvantage. With reduced incidence of sunlight, radial development of the crowns is restricted and growth is slowed down.

In fact, stagnation may take place, especially in species showing little differentiation in height growth. It is the aim of the silviculturist to make the most of early competition while this is advantageous and to remove it when it hinders stand development — in other words, to distribute the total productive capacity of the site among the optimum number of trees to obtain the desired results. This is usually accomplished by thinning operations.*Thinnings.* Cuttings made in a fully stocked to overstocked immature stand in order to increase the rate of growth of the remaining trees are known as "thinnings." In general, these are begun in the late sapling stage and are continued periodically until the stand matures: they are usually of only moderate intensity until rapid height growth is completed, the later thinnings being more severe to accelerate diameter growth.

The total annual growth of an area is not materially changed by thinning, but it is confined to fewer trees whose individual increments are thereby increased. As a result, the trees reach useful sizes at earlier ages, which is particularly important when logs of large size are desired for sawing into lumber. Too heavy a thinning, however, may leave insufficient trees for full use of the productive capacity of the site.The economic feasibility of thinning operations is largely determined by the availability of markets for small-sized material, although in some cases the improved growth of the residual trees may justify the operation.*Other Cultural Treatments.*

To promote more desirable species and better formed trees and generally to improve the quality of the stand, other cultural treatments may be used, among them cleanings, liberation cuttings and improvement cuttings.

- *Cleaning.* An operation in a young stand not past the sapling stage:
  - To free small trees from weeds, vines, or sod-forming grasses;
  - To provide better growing conditions by liberating crop trees from other individuals of similar age but of less desirable species or form that are overtopping or likely to overtop them.
- *Liberation Cutting.* The release of young trees not past the sapling stage from competition with older trees that are overtopping them.
- *Improvement Cutting.* A cutting made in a stand past the sapling stage to improve its composition and character by removing trees that are less desirable with respect to species, shape, or main crown canopy.

*Pruning.* Knots in lumber are the result of branch growth; the fewer and smaller the branches, the fewer and smaller the knots and the higher the grade of lumber produced. In a young stand of proper density the lower branches are often small in diameter and die at a fairly early age.

The forester may assist in the production of knot-free wood by removing the dead and dying branches from the bole of the tree with axe or knife. Care must be taken, however, not to injure the tree nor to reduce unduly the living portion of the crown.

A comparison of the cost of pruning with the relative value of knot-free wood will determine the feasibility of this treatment.

## DIFFERENT TRENDS IN RECENT SILVICULTURE

Another grouping can be made according to two different trends in recent silviculture, or two different attitudes towards the forest and these attitudes will surely be of more critical importance in the future than any particular system of forestry. The attitudes are not always well expressed, but in their clearest form, one of them can be described as faith in artificial forest, the other as faith in natural forest.

On the one hand, there are those who believe that the forest should be treated as a plantation producing a maximum amount of the type of wood that is most profitable to man, the pace of production being forced by human action; on the other hand, those who think that the best result in the long run will be obtained if natural conditions in the forest are disturbed as little as possible and the forest, while encouraged by the co-operating hand of the silviculturist, sets the pace of timber production.

The difference some what resembles the contrast in agriculture between those who believe in "organic farming" and those who have what Sir Albert Howard calls the "N K P mentality" and place their faith in forced feeding with commercially produced chemical fertilizers. These attitudes towards the forest are found in every country, but the history of German forestry provides the clearest examples and the longest record.

Coppice, clear-cutting and selection-cutting have been practiced ever since the Middle Ages in the forests of Germany and western Europe generally, locally more or less regulated at certain times but more commonly not. Unregulated selective cutting in high forest was widespread, but it was the wrong kind of selection: the best trees were cut out, leaving the poorer specimens to regenerate the forest.

Some of the forests were excessively grazed by livestock; the custom of gathering fallen material for fuel and the raking up of leaves for bedding the livestock in winter denuded many forests of their litter; and yet other forests, because of the passion for hunting among the nobility, becamc overstocked with game, causing great damage to the forest seedlings.

The result was a general deterioration of the forest, which by the end of the eighteenth century had become so serious that strong remedial measures were initiated. Heinrich Cotta, 6 Director of the Forest Academy at Tharandt, suggested that the only way of rehabilitating the forest was by clear-cutting and making a fresh start, helping the regeneration by the planting of trees, if necessary. This method became common in the forests of Saxony and was adopted in many other regions as well. It should be noted that the proposal was remedial in purpose, not intended as a permanent system and Cotta himself warned against the danger of monoculture. But that is what the method turned out to be.

It became standard practice in the nineteenth century and early part of the twentieth to clear-cut, plant and harvest on short rotations and the upshot was that many forests became dominated by even-aged, nearly pure stands of spruce or pine, while fir and beech, oak and other deciduous trees well-nigh disappeared. The dominant motive was profit: spruce or pine, worked on short rotations for lumber, pulp, mine timber and the like, gave-for a while-the highest and quickest economic returns. This was the earliest and one of the clearest expressions of the first of the two attitudes, faith in artificial forest. As Franz Heske says, wood changed from a carefully rationed essential material to an ordinary commodity, the production of which was governed primarily by financial considerations.

But signs that all was not well began to appear: stunted growth of trees, yellowish needles, increasing growth of lichens and peat, decrease in grass, formation of hardpan in some soils; and it also became apparent that the trees suffered increasing damage from wind, frost, fungus and insects. A series of thorough investigations of forest conditions in Saxony, undertaken by E. Wiedemann in the 1920's, revealed marked soil deterioration and a loss in the average annual growth rate, which according to estimates quoted by R. S. Troup amounted to more than 9 cubic feet per acre over a 100-year period, or a total loss of 3 1/2 billion cubic feet over the whole of the state forests in Saxony. Unsatisfactory results in managed forests have also been observed in other regions.

Reports from the large planted pine forest of the Landes in southwestern France indicate that the heavy losses of recent years have resulted not only from catastrophic fires, windfalls and attack by beetles, but also from the more fundamental cause of soil deterioration.

After about 100 years of intensive management, described by G. Roux as "absolute monoculture," the trees are less healthy, the annual increment has dropped, the yield of turpentine has decreased and natural regeneration is almost impossible to obtain; seed cones have long been imported. J. L. Arend estimates that the area underlain with hardpan in the region has increased by approximately one half.

One of the proposed remedies is the introduction of broadleaf species to mix with the pines. Anders Holmgren, after fifty years of work in the forests of Swedish Norrland, during which time he has given much of his attention to the problems of regeneration and methods of cutting, recommends greater reliance on natural forest and natural reproduction, advises against clearcutting with the coulisse method in the mountains and the far north and suggests that, if the method must be used, the areas cut should not be larger than 25 to 35 acres. The experience of Saxony does not prove, as many foresters have pointed out, that clear-cutting alone was responsible for the deterioration, nor that pure stands are necessarily bad: some forests naturally consist of nearly pure stands and adverse results have also been observed in mixed forests managed with different systems.

The reasons for the unsatisfactory outcome were many and complex, but certainly the fundamental cause was the forced pace dictated by financial considerations. It is not likely that any forest soil, whether supporting pure or mixed stands, can long stand up under intensive short-rotation cropping.

Not all foresters approved of the system. Strong protests were made, particularly by Karl Gayer, who advocated the reestablishment of forestry operations based upon and not contrary to, natural laws.

Many followed Gayer's lead and a reaction set in against clear-cutting and monoculture, reaching an opposite extreme in the idea of Dauerwald, or forest with continuous cover and selective-cutting in such a manner that "the forest hardly notices it" Clear-cutting and artificial regeneration are still practiced in many regions, but the present trend in European silviculture is in another direction. E. W. Jones summarises his survey of present-day ideas among European foresters by saying that opinions preponderantly favour mixture of species, natural regeneration, constant preservation of the forest canopy and flexible cutting methods and stress particularly the importance of giving careful attention to soil and site conditions in reaching decisions on silvicultural policy for a given region. That forestry is essentially an economic operation is not overlooked; on the contrary, the belief is that these methods of cooperating with the bent of nature will in the long run give the highest returns and at the same time preserve the health and beauty of the forest. It is interesting to note, as an indication of the trend, that Joseph Köstler concludes a recent major work by reminding us that "forestry is also landscape care," and leaves the reader with a quotation from Goethe: "We must eavesdrop on Nature to learn her ways, lest we coerce her into obstinacy by procedures against her will" Here is the clearest expression of the second attitude, faith in natural forest.

The trend of European silviculture may be towards natural forest, but whether the industry as a whole will follow suit is another question. There is no doubt about the persistence of strong faith in artificial forest and forced pace in countries outside of Europe. The new fast-growing pine forests in Chile,

New Zealand, Australia, Kenya and South Africa are essentially monoculture plantations, with some of the rotations as short as thirty years. The technique developed in South Africa, as described by W. E. Hiley, consists of early and heavy thinning according to mathematical rules and "has the object of producing the greatest value of timber in as short a time as possible and the prescriptions are based on strictly economic principles" No better definition of nineteenth-century forestry in Saxony could be found, but whether the result will be the same only future generations will know. Hiley concludes that it is too early to pronounce judgement on the principles of this "daring experiment"

The large industrial forests in the southeastern part of the United States, where pines have always been abundant, are becoming dominated by softwoods at the expense of hardwoods because of the emphasis on planting pine for industrial purposes. Southern forestry, in fact, seems to be destined for monoculture.

The prevailing attitude is indicated in speeches made at the meeting of the Southern Pulpwood Conservation Association in February 1954. As reported in American Forests, one of the speakers said: "Put every acre to work -- that's good management, make every acre work harder -- that's applied research, make every acre grow better trees faster-that's basic research" The speakers also questioned the wisdom of growing high-grade saw timber on 80-year rotations in the national forests of the South, where there are so many pulpmills. The planting of trees is certainly a commendable practice and no one can object to improvements in management and yield, but it is obvious that here we are again on the old Saxon road. The intention is to make the acres do what we bid. It will also be observed that the statement of the speaker reflects the widespread confusion over what constitutes basic research. Not only is the task of basic research assigned, but the expected result is laid down.

One can imagine the feeling of a research scientist: produce better trees, or else! In the rugged terrain of the dense Douglas fir forest in the Pacific Northwest the method now almost universally employed and approved by virtually every forester familiar with the region, is clear-cutting of staggered blocks, sometimes called "area selection" There is no standard size of the blocks to be logged, for conditions vary from place to place, but in conversation with foresters having lifetime experience with Douglas fir one hears the opinion that, by and large, the blocks ought not to exceed 30 or 40 acres.

It may be recalled that Holmgren made the same suggestion for northern Sweden. But the economic advantage of cutting larger blocks is great and clear-cut areas of 300 acres are common. It can hardly be said that the long-run effect on the forest of such large blocks is known, but many foresters suspect that it will not be good. The size of the blocks is indicative of an attitude that is determined more by financial interest than by silviculture.

We are clearly moving in a direction opposite to that suggested by Goethe. We intend to coerce nature, hoping that she will not respond with obstinacy, but there is plenty of evidence that she will and none that she will not. The claim that good management can cause-or force -- the American forest soil to yield twice as much wood as now on a sustained basis has almost become sacrosanct doctrine. Some speak of increasing the yield even more. We know that forestry has improved remarkably in recent years, we know that good management can raise the annual growth, but we do not know, we only hope, that the total annual production can be doubled and permanently maintained at that level. The science of genetics has come to the fore during the last half century and may conceivably have a tremendous effect on the forests of the future. The renewed interest in natural forest during Karl Gayer's time was accompanied by a reawakening curiosity about the nature of the forest, the ironic result of which may be the end of both the concept and the existence of natural forest. Like silviculture, modern forest genetics has its precursors.

Duffield points out that as early as the 1780's a German forester by the name of Wangenheim was aware of hereditary strains in forest trees and recommended the planting of seeds from superior individuals and that Patrick Matthew, a British forester, in 1831 proposed a theory of natural selection derived from his experience with raising trees. In more recent years it has been discovered that there are local races within the species of trees and that a number of traits are hereditary, such as form of trunk, type of branching, rate of growth, vigour of seeding. Applied forest genetics is based on these discoveries. It is no longer a question merely of planting seeds but of planting seeds from the best trees.

Several thousand "elite" trees have been registered in the forests of Scandinavia and Finland, seeds are regularly collected from these protected individuals -- with the aid of specially constructed steel ladders erected from movable platforms, making it possible to harvest the seeds without harming the tree -- and careful records are kept of the progeny. C. Syrach-Larsen began to establish seed tree orchards in Denmark during the 1930's for the purpose of raising scions of superior trees and to carry out cross-pollination experiments. In Sweden pines from the north, characterised by slow growth but highquality wood, have been crossed with faster growing southern stock in the hope of combining the best qualities of both parents.

Experiments with X-ray treatment to achieve polyploid varieties have produced a "giant aspen" having twice the ordinary number of chromosomes and maturing in 30 years. J. W. Duffield and F. I. Righter report about two dozen new pine varieties produced at the experiment station near Placerville in California, many of which show decided hybrid vigour, grow faster and are more resistant to cold, insects and diseases than either of the parent species.

A cross between eastern white pine and Himalayan white pine is watched with particular interest, for this hybrid not only outgrows eastern white pine but has shown itself resistant to blister rust in a rigourous exposure test that now has been in progress for nearly a decade.

Perhaps we are on the threshold of a period that future historians will refer to as the time when forest trees were domesticated and natural forests passed away. It is to be expected, if the domestication succeeds, that the character of the future forests will depend on present trends in applied forest genetics. While all geneticists work in the same general direction with controlled breeding experiments, there seem to be different opinions about the advantage of exotics over native species, some placing most of the faith in elite indigenous trees, others expecting more from introduced species, or from a combination of exotics and superior native stock.

## SILVICULTURAL SYSTEMS

As a result of past study and experience, particularly in northwestern Europe, there has been developed a series of silvicultural systems, each incorporating the techniques that are required in the silvicultural management of a specific type of forest to maintain its distinctive form and maximum productivity.

While these systems are based primarily on the method of regeneration (whether by coppice or seeding and with reference to the form of cutting), they are also concerned with problems affecting the protection and tending of the forest and the harvesting of the crop and its economic utilisation. Where such systems have been evolved, they will form an essential element in the plan of forest management.

## IMPROVEMENT OF THE FOREST STOCK INTENSIVE SILVICULTURE

The principal trend is no doubt in the direction of combination. On the whole, European experience with introduced species has not been particularly happy; and J. S. Boyce, in a thought-provoking article, describes the risks of spreading diseases and insect pests by introducing foreign trees, warns against placing too much faith in them and concludes by saying that "exotics are not all foredoomed to failure, but for every exotic the chance of failure appears to be much greater than the chance for success" The "combination" idea is expressed by Syrach-Larsen in the following statement, although, if anything, he seems to favour exotics:

"Broadly speaking, for improvement of the forest stock intensive silviculture places most reliance on the introduction of exotics and on selection of the best indigenous trees" He also says: "To succeed to any appreciable extent we must soon approach the stage when it becomes obvious how much more profitable it will be to create rapid-growing pure plantations on favourably situated areas"

This is artificial forest with a vengeance. The image conjured up resembles a field of hybrid corn: super-trees, single cropping and fast rotations. Perhaps the forest operator of the future, like the farmer now, will have to spend a large part of the income for fertilizers. In line with this trend a representative of the fertilizer industry attended the Fourth American Forest Congress in 1953 and suggested a programme of spreading fertilizers from airplanes on forest soils, thereby "permitting the cutting of trees perhaps ten years sooner"

The problems in nature facing the forester are many and complex: fire, windfalls, frost damage, insects and fungi, but surely the most critical long-run problem in any silvicultural system, whether of the traditional type or the new, is the maintenance of a fertile and healthy forest soil; for if the soil fails, everything fails.

Fortunately the foresters, at least in some regions, are well aware of the problem. C. W. Scott stresses the importance of watching the soil in the new Chilean plantations of Monterey pine and André Aubréville has given the same advice to the managers of eucalyptus groves in Brazil. Most of the larger companies in the forest industries of the Pacific Northwest have had soils experts on their staffs for several years.

Watching the soil is one thing, but providing a remedy if it should turn out to be deteriorating is another. It would hardly be economically feasible, even if it were physically possible, to raise forest trees with the aid of fertilizers, as we raise many of our crops. My own belief is that the best promise for the future lies in another direction, namely natural forest, but we do not seem to be going that way.

In a purely technical sense, silviculture in the forest is perhaps as bright a promise as technology at the mill, but both run into the dilemma of the small-ownership and economic problems not of their making.

Excellent results have been obtained by application of the Dauerwald idea in some of the European forests, but Dauerwald, which requires, so to say, Dauerpflege or continuous and careful attention to details and much hand labour, is, for economic reasons, impossible in most of the world and certainly impossible on both small and large forest holdings in North America.

Every forest operator knows that he cannot always follow even the best advice of the silviculturist, for he would go broke if he tried. Clear-cut patches in the Pacific Northwest on which Douglas-fir seedlings should be coming up are in many regions taken over by alder or choked with brushy vegetation; some of the young stands need thinning; and many trees should be pruned for best results.

Only the larger companies are able to perform improvement operations of this type, for money spent on pruning, thinning, or fighting brush might not be returned until 75 or 100 years have elapsed and under present circumstances most of the operators are unable to finance such long-term investments.

The result is that much of the forest is growing up unattended and the land is not producing so much good timber as it might. According to the Annual Report of the Forest Experiment Station at Portland in 1952, the use of pulp chips obtained from mill waste in Oregon and Washington has meant an annual salvaging of material in recent years equivalent to several hundred million board feet of logs, a very good saving, but at the same time the practice has had a detrimental effect on the market for pulpwood obtained by thinning, making good forest management more difficult. Thus, what is good economy at the mill is not always good silviculture in the forest.

## CULTURAL PROMISES AND BLOCKS

The outlook can be summarised as follows: Although shrinking, the area of the accessible forest is large enough and our technical skill, even though problems remain, is now good enough to supply all mankind with a reasonably adequate amount of wood in the near future, provided the skill be freely applied over the area and the yield of the forest be equitably distributed. But deeply entrenched cultural factors interfere with both the application of the skill and the distribution of the yield. This is the root of the forestry problem and the most difficult part of it.

We know how to change the number of chromosomes in aspen and how to produce pines that are resistant to blister rust, but we are less successful in dealing with ingrained human habits and the blister rust on our cultural institutions. Some people want to save the world by changing the old ways of life, others want to save it by resisting change, but it is not clear that the first group really understands what is involved in its proposal, nor the second what it is resisting.

Culture history can be called a history of change, but it is also a history of the persistence of customary ways and we still have much to learn about both. We hear much today about the need of basic research and if anything is in need of close scrutiny it is the man-earth relationship, not only for what are called academic reasons but also for practical purposes. The future of the forest depends on such research, for between man and the timber is always a third factor, human culture. Some of the cultural factors affecting forestry are promises, others are blocks and both are so numerous and complex that only a few examples can be considered here.

It is a promising sign that more people than ever seem to be aware of the fact that there is a forest problem. Throughout the world, movements are afoot to save the forest, to plant trees, to introduce more efficient uses of wood and to foster a better and more widespread understanding of the importance of the forest in the protection of soil and water.

There is, however, an element of risk involved. Silviculture can never be in a hurry, but popular movements sometimes are and the danger is that mistakes may be made because of the hurry. An error in the forest may take a century or two to correct, if it can be corrected at all. We have blister rust in the United States today because of the mistake of introducing infected white-pine seedlings.

Mention has already been made of economic difficulties that encumber technology and silviculture and of the fact that both are caught in the dilemma of small ownership. This dilemma has been resolved -- or removed, rather -- in the countries where the state holds all the forest land. But this is not to say that the forest problem has been solved and systems of land tenure are changing so fast at the present time in many regions that not much can be said about what may come out of the flux. In most of the Western world, however, the problem of small ownership remains. It is regarded as a problem by foresters because it is generally on the small lots that forest management is poorest and most difficult to improve.

Even in the Scandinavian countries and Finland, where the populations are relatively small, culturally homogeneous and strongly forest-minded, management on many small properties leaves something to be desired. Franz Heske refers to the small private woodland -- Bauernbusch -- as the "child of sorrows" of German forestry; and the consensus among foresters in the United States is that one of the most critical problems is how to improve management on the small holdings. It is not a little problem. One half of the land classified as commercial forest in the United States -- the half that is generally most accessible and potentially productive -- is owned by some four million persons, holding on the average about sixty acres each.

Most of this land is in the form of wood lots owned by three million farmers and the rest is held by a million non-farmers, many of whom are absentee owners. Numerous difficulties stand in the way of better forest management on the wood lots: their uneconomically small size for the purpose of timber production; the commonly frequent change of ownership; the lack of capital and experience in forestry; sometimes the lack of interest, for the farmer is primarily occupied with raising food crops, not trees; and problems of marketing. A promising development is the formation of associations of wood-lot owners for dealing with these problems, for example the "Otsego Forest Products Cooperative Association" in New York, "Connwood, Inc" in Connecticut, "The Shelton Cooperative Sustained-Yield Unit" in western Washington and others. The technical assistance of some two or three hundred government foresters is available to wood-lot owners and a number of industry-sponsored organisations are also encouraging better forestry on the small holding, such as "Trees for Tomorrow" and "Cash Crops from Your Woods" There must be several thousand wood-lot owners now belonging to associations and actively engaged in improving forest management on their lots, but that is still a long way from four million members.

Improvement is a good thing, but we have fallen into a rut in talking about the imperative necessity of raising the timber production on the farm lots and we need to give some thought to another side of the question, the farmer's side. There is wisdom in what Franz Heske says: "The small woodlot must not be judged by its yield of timber, for its importance lies in other values: fuelwood, protection of soil and water and shelter from the wind. The woodlot will probably never be an important source of supply for the timber market" This comment also applies to North America. I suspect that many an owner likes his wood lot as it is and does not think of it as commercial forest land, but rather as a place for getting some firewood and fence posts, doing a little hunting and fishing, or just walking in the woods. "There are some who can live without wild things," says Aldo Leopold, "and some who cannot" Wood-lot owners wishing to improve timber management on their lands should certainly be encouraged, but we have no right to force the farmer into forestry and the number of people who like their woods just as they are may be larger than we think.

## IMPRESSION OF THE TIMBER RESOURCE

It is unrealistic and gives a false impression of the timber resource in the United States to classify some 200 million acres of wood lots as commercial forest land, when in fact many of them are not and perhaps never will be. It might be wiser to count them under some other heading in the census and face the timber problem on the basis of what we know is commercial forest land. Likewise, it would be sounder to think of the areas given over to shifting cultivation in the tropical regions as farm land, which they are, instead of calling them potential timberland. We have as yet no adequate practical way of changing either the system of shifting agriculture in the tropics or the wood lots in the midlatitudes. Both are firmly rooted cultural institutions.

Some of the problems are perhaps no more than vexatious, but nevertheless tend to hinder good management. In the United States, for example, most of the land is mapped according to the well-known township and range system, which is a good system in level country, but its suitability in the rough terrain of the western forest lands can be questioned. The sections lie over the country like squares on a checkerboard and property lines run straight across mountains, ridges, valleys and rivers without regard to the lay of the land.

Ownerships are chaotically mixed up and the chaos is worst precisely where the best timber grows--that is, in the redwood belt of California and the Douglas-fir region of Oregon and Washington. The® system naturally creates difficulties with access roads and causes other trouble. It is a basic principle in forestry that good management is facilitated by contiguous holdings and rendered difficult by scattered property. The intricate pattern of ownership may in time be unscrambled even though the process of unscrambling does not seem to move very fast, but there is little hope of ever changing the checkerboard system, for it is an institution so thoroughly fixed in American tradition that no one even talks about changing it.

Another problem is forest taxation, which in one form or another has probably troubled forestry in every country. In the United States, at least, the manner of taxing timber has in the past worked against rather than helped good forest management and in many states of the union it still does. Property taxes on growing timber, paid annually and increasing as the trees approach maturity, have forced many owners to clear-cut prematurely in order to salvage some of the value of the timber before it is all eaten up by taxes. New Hampshire has introduced a "yield tax" calculated on the anticipated value of the timber at the time when it will be cut and furthermore allows a rebate to owners who manage their forest lands according to certain standards. Several other states have also begun to change their tax systems, but most have not. Other factors having an adverse effect on forest management in the United States are the banking and insurance regulations, which make it difficult and in some instances virtually impossible to insure forest land or to obtain a loan with standing timber as security.

The tradition of the virgin forest, or the lure of new timberland to be opened for utilisation, is as old as forestry itself and is still very strong. The timber in the lands of the western Mediterranean probably looked as inexhaustible to the ancient Greeks and Phoenicians as did the forests of Germany to Caesar and Tacitus, or the American forests to the European colonists, or the tropical forests to some people today.

This tradition is not helping us to settle down to good management, for we seem unable to face the problem as long as there are virgin forests over the hill waiting to be opened up and made "fully productive" There is truth in the saying that good forestry is the child of necessity. According to Theophrastus, the ancient rulers of Cyprus began to conserve their more accessible cedar groves only when it became too expensive to transport timber from the interior. The forest regulations that came into existence in western Europe during the late Middle Ages were in large measure stimulated by the fear of wood famine. In the Pacific Northwest, as a veteran forester once told me with a trace of bitterness in his voice, the timber buccaneers began to practice conservation after they had slashed their way across the continent to the coast and could go no farther. However, we should not visit the sins of the fathers on the present generation of lumbermen, for many of them are helping to create a better tradition. It is significant that the leading timber companies now regard the forest land they hold--the soil--as their most valuable possession, not the timber standing on it at the present moment.

The distinction is important. But the emergence of the new tradition does not mean that the old yearning for virgin forests has ceased. On the contrary, the pressure for opening more timberland continues as strong as ever. One can scarcely read anything dealing with forestry, whether it is a government report, industry-sponsored publication, or proceedings of a forest convention, without encountering the claim that one of the greatest needs is more access roads into the mature and overmature virgin forests. It is true that more timberland will gradually be needed in order to replace forest areas lost to other forms of land use in the future, but it would also seem to be true wisdom not to open the virgin forests any faster than we must.

There are several good reasons for a cautious approach. An understanding of how trees reproduce is fundamental to silviculture, but very little is known about the fruiting habit or, in fact, about the whole regeneration process of most of the species of trees in the tropical forests. P. W. Richards suggests that we need more basic research into the nature of the tropical forest before we try to manage it. Even in the midlatitudes, as J. W. Duffield points out, the phenomenon of seed years is not yet fully understood. Forest trees have always suffered from diseases and insects, but in recent years the attacks of these enemies have developed into dangerous epidemics that threaten the very existence of some species of trees. W. B. Greeley reminded the Mid-Century Conference on Resources for the Future that in trying to control insects and diseases we are seriously handicapped by a lack of fundamental knowledge.

Furthermore, we do not know to what extent these epidemics are the result of our opening of the forests. It is not inconceivable that opening the virgin forests might be tantamount to inviting a more intensive attack upon them by insects and diseases. The truth of the matter is that modern silviculture and forest management are still in the experimental stage and it seems unwise to experiment with our last reserves of virgin timber, particularly since we have an experimental field closer at hand.

If it is true, as has often been claimed, that good management can double the yield of the forest and if we think we are skilled in managing timberland, it would seem only common sense not to go into the virgin forests until we have proved our skill by first putting in order some 75 or 100 million acres of already accessible but unproductive and poorly stocked forest lands that lie strewn over the United States in a cut-over, burned-over and generally wretched condition as reminders of the fact that a short while ago we did not manage well. "We should recognise the importance of placing heavy investments on denuded lands or poor quality stands that occupy the best quality lands. This is more important for long-run future production than rushing into our few remaining virgin stands" One would expect that this sensible and important suggestion should have stimulated a lively discussion among the letter writers, or for that matter on the editorial page, but to my knowledge the response has been a dead silence. The prevalent mode of thinking seems to be that we must open all the timberlands as soon as possible and then, presumably, we shall put the whole house of forestry in order.

The real drive behind the demand for opening the reserves, is the old yearning for virgin timber, but the tradition has been rationalized in recent times by saying that it is not good conservation to leave the virgin forests alone. The argument is that wood is "wasted" in the mature forests because old trees are dying. It is a specious argument, based on the old mistaken idea that the standing timber constitutes the resource, whereas the real resource is the land that produces the timber.

It is true that trees are dying in the mature forests, but trees are also growing up, as they have been doing for millions of years. There will be timber and what is more important, there will be timberland in the virgin forests even if we do not get into them for a long time to come. We are not wasting any of the true resource--timberland-by staying out of the virgin forests until we actually need them, but we will be wasting a part of the resource unless we soon concentrate a major effort on restoring the lands that are denuded or poorly stocked because of former mismanagement.

It is not possible to place the cultural promises and blocks in a balance free from personal bias in order to see which outweighs the other. The chances are that the inclination of the person doing the weighing would decide the outcome, so that one might see a bright future ahead and another would not. Cultural institutions are not fixed forever; they change and it may be that the culture blocks will be removed in time, but the evidence at hand does not suggest to me that we have been very successful in this removal.

## SILVICULTURE AND MANAGEMENT OF FOREST

Although silviculture and management may be looked upon as two branches of forestry, the one mainly technical and the other economic, the two are interdependent and must always be coordinated. While silviculture may establish the ideal possibility of production within a forest, management determines the degree to which this can be realized through regulation and within the bounds of good business practice. Forest management may be defined as the application of business methods and technical forestry principles to the operation of a forest property.

Compared with other crops, a forest requires a relatively long time between its establishment and its harvesting. Revenue from an unmanaged forest may accrue only at fairly long intervals. Managed forests are organised to ensure a sustained yield of the forest crop in perpetuity. The income should include a reasonable profit on the investment, annually if possible.

Sustained yield of timber depends upon the systematic reproduction of a crop as it is harvested. It also requires that the timber be cut annually or periodically in such quantities that a continuous and fairly uniform yield will be provided throughout the rotation (the period required to grow a crop from seed to maturity).

The principle of sustained-yield management may be illustrated with reference to a theoretical forest, usually termed the normal forest, in which the stands are stocked with the maximum amount of wood that the site will produce and from which the same amount of wood product may be harvested annually forever-namely, the amount that grows annually on the whole acreage.

Consider, for example, an even-aged forest that is to be harvested when fifty years of age. To have fifty-year-old timber every year, there must be a series of fifty stands of, respectively, one, two, three and up to fifty years of age. Each year one stand becomes mature for harvesting. This stand in volume represents its own increment or growth for fifty years; also it is equal to the growth of one year on the fifty stands making up the whole forest. Accordingly, each cut is equal in amount to the current year's growth. It is assumed that each stand is reproduced as cut and that all the immature stands receive the silvicultural treatment necessary for most satisfactory growth.

An alternative method, for an uneven-aged forest, is to cut over the entire area periodically, removing at each cut only the volume corresponding to the growth during the period between successive cuts.

In either case, one has permanent sustained yield from a forest characterised by normal distribution of age classes, normal increment and normal growing stock. There is, of course, no such forest. Every forest is abnormal, having either excess or deficiency of area in the older age classes and the opposite in younger ones and in general a deficiency, to a greater or less extent, in volume per acre in the stands of different ages.

The abnormalities in age-class distribution can be removed only by adjustments in the areas cut over in successive years, while deficiencies in volume, caused mainly by inadequate reproduction, can be made up through improvement in silvicultural practices. The normal forest is an ideal, a standard towards which forests should gradually be developed.

Such intensive management requires accurate information about tree species and sizes, total and merchantable volume, age-class distribution, growth conditions, site classes and rotation ages, as well as about such factors as topography and drainage, which affect the efficiency of operations. Such information is obtained through enumeration or inventory surveys, supplemented and improved in recent years by aerial photographs.

On the basis of this information, a management plan or working plan is evolved, which sets down the objectives of management and how they are to be attained and allocates the kind and amount of timber to be cut each year or period of years, at least for the first part of the rotation.

## ENGINEERING OF LOGGING

Logging engineering is concerned essentially with the felling and removal of timber from a forest area. Even where the sole purpose is liquidation of the resource, logging techniques may be developed to a high degree of efficiency, as they have a direct and important bearing on the cost of extraction. In forests under sustained-yield management, logging methods must be developed to meet the requirements of the management plan as well as to maintain operational efficiency.

Logging everywhere presents many technical problems, but particularly where the forests are difficult of access, distances are great and the topography is rugged. In many forest regions logs are transported long distances and at relatively low cost by natural water systems and in the northern forests winter snows often facilitate the hauling of forest products.

From the point of view of forest conservation, perhaps the most important logging objectives are:

- To prevent damage to the advance growth and the residual stand,
- To help prepare the site for future tree crops (*e.g.*, by soil scarification),
- To ensure minimum waste of wood in the trees cut.

In recent years mechanical equipment has been increasingly employed in both the felling of trees and the transportation of logs.

## WOOD UTILISATION IN FOREST

While it is not usual for those responsible for the growing and harvesting of timber crops to be directly concerned with their conversion and manufacture into usable products, there should be a close liaison between the two operations of forest production and wood utilisation. Wood has inherent qualities that make it particularly suitable for many purposes—good strength in relation to weight, pleasing appearance, insulation against heat and sound, ease of fastening with nails, dowels, screws, or glue and case of working with relatively simple tools. However, wood is by no means a homogeneous material: it differs from one species to another in structure and physical properties and hence in the use for which it is best suited.

Within environmental limitations it is desirable to grow the species that will best meet the requirements of local industry and available export markets. Furthermore, to make the fullest use of the timber crop at all stages of development and to minimize logging and manufacturing waste, it is important to co-ordinate as closely as possible forest operations and the wood-using industries and also to bring about a closer integration of the wood-using industries themselves — as, for example, the sawmill and pulp industries.

### Forest Policy

In many countries forests are of vital importance to the over-all economy, not only in relation to industrial development, but also in providing benefits which may be difficult to evaluate in monetary terms. For this reason and also because of the long-term nature of the forest enterprise, most countries seek to ensure the perpetuation of their forests and to safeguard the interests of present and future generations by placing the forests under some degree of government control.

In some countries the government retains ownership of a large proportion of the forested land and leases it to the wood-using industries for timber extraction: provision may be made for some degree of forest management by regulation or through agreement. Other countries have granted or sold much of their forest land to private interests, though considerable areas have sometimes been retained by the state to provide for timber reserves, recreational facilities and the protection of drainage areas; in some cases there is considerable public control of forest management on the freehold lands.

Elsewhere long-term forest management is entirely neglected and forest policies, if any exist, may be concerned only with the method and intensity of exploitation.

Progress in the development and practice of forestry varies widely throughout the world. While forestry is well advanced in much of Europe and in certain tropical regions, only a small proportion of the accessible forests of the world is under intensive management today. In some regions man still depends on the forests for his food and shelter and contributes little towards their development; in many countries man long ago destroyed the forests and has done little to restore them; elsewhere uncontrolled forest exploitation is still in evidence and proper protection and management lie in the future.

In North America we are still in an early stage of forest conservation. In recent years, however, we have begun to see the forests as an integral part of our economic structure and to take steps against the destructive exploitation that has threatened to exhaust them.

# 3

# The Value of Forest Ecosystems

## NUTRIENT CYCLING

In ecology, a biogeochemical cycle/ nutrient cycle is a circuit or pathway by which a chemical element or molecule moves through both biotic ("bio-") and abiotic ("geo-") compartments of an ecosystem. In effect, the element is recycled, although in some such cycles there may be places (called "sinks") where the element is accumulated or held for a long period of time.

All chemical elements occurring in organisms are part of biogeochemical cycles. In addition to being a part of living organisms, these chemical elements also cycle through abiotic factors of ecosystems such as water (hydrosphere), land (lithosphere), and the air (atmosphere); the living factors of the planet can be referred to collectively as the biosphere. All the chemicals, nutrients, or elements—such as carbon, nitrogen, oxygen, phosphorus—used in ecosystems by living organisms operate on a closed system, which refers to the fact that these chemicals are recycled instead of being lost and replenished constantly such as in an open system. The energy of an ecosystem occurs on an open system; the sun constantly gives the planet energy in the form of light while it is eventually used and lost in the form of heat throughout the trophic levels of a food web.

The Earth does not constantly receive more chemicals as it receives light. The Earth only has those chemicals that were formed in the creation of the Earth, and the only way to obtain more chemicals or nutrients is from occasional meteorites from outer space that contain those elements. Because chemicals operate on a closed system and cannot be lost and replenished like energy can, these chemicals must be recycled throughout all of Earth's processes that use those chemicals or elements. These cycles include both the living biosphere, and the non-living lithosphere, atmosphere, and hydrosphere. The term "biogeochemical" takes its prefixes from these cycles: Bio refers to the biosphere. Geo refers collectively to the lithosphere, atmosphere, and hydrosphere. Chemical, of course, refers to the chemicals that go through the cycle.

The chemicals are sometimes held for long periods of time in one place. This place is called a reservoir, which, for example, includes such things as coal deposits that are storing carbon for a long period of time. When chemicals are held for only short periods of time, they are being held in exchange pools. Generally, reservoirs are abiotic factors while exchange pools are biotic factors. Examples of exchange pools include plants and animals, which temporarily use carbon in their systems and release it back into the air or wherever. Carbon is held for a relatively short time in plants and animals when compared to coal deposits. The amount of time that a chemical is held in one place is called its residence.

The most well-known and important biogeochemical cycles, for example, include the carbon cycle, the nitrogen cycle, the oxygen cycle, the phosphorus cycle, and the water cycle.

Biogeochemical cycles always involve equilibrium states: a balance in the cycling of the element between compartments. However, overall balance may involve compartments distributed on a global scale.

Biogeochemical cycles of particular interest in ecology are:

- Nitrogen cycle
- Oxygen cycle
- Carbon cycle
- Phosphorus cycle
- Sulphur cycle
- Water cycle
- Hydrogen cycle

## NITROGEN CYCLE

The nitrogen cycle is a much more complicated biogeochemical cycle but also cycles through living parts and non-living parts including the water, land, and air. Nitrogen is a very important molecule in that it is part of both proteins, present in the composition of the amino acids that make up proteins, as well as nucleic acids such as DNA and RNA, present in nitrogenous bases. The largest reservoir of nitrogen is the atmosphere, in which about 78 per cent of nitrogen is contained as nitrogen gas (N2). Nitrogen gas is "fixed," in a process called nitrogen fixation. Nitrogen fixation combines nitrogen with oxygen to create nitrates (NO3).

Nitrates can then be used by plants or animals (which eat plants or eat animals that have eaten plants). Nitrogen can be fixed either by lightning, industrial methods (such as for fertilizer), in free nitrogen-fixing bacteria in the soil, as well as in nitrogen-fixing bacteria present in roots of legumes (such as rhizobium). Nitrogen-fixing bacteria use certain enzymes that are capable of fixing nitrogen gas into nitrates and include free bacteria in soil, symbiotic bacteria in legumes, and also cyanobacteria, or blue-green algae, in water.

After being used by plants and animals, nitrogen is then disposed of in decay and wastes. Detritivores and decomposers decompose the detritius from plants and animals, nitrogen is changed into ammonia, or nitrogen with 3 hydrogen atoms (NH3). Ammonia is toxic and cannot be used by plants or animals, but nitrite bacteria present in the soil can take ammonia and turn it into nitrite, nitrogen with two oxygen atoms (NO2). Although nitrite is also unusable by most plants and animals, nitrate bacteria changes nitrites back into nitrates, usable by plants and animals. Some nitrates are also converted back into nitrogen gas through the process of denitrification, which is the opposite of nitrogen-fixing, also called nitrification. Certain denitrifying bacteria are responsible for this.

**oxygen cycle**

The oxygen cycle is the biogeochemical cycle that describes the movement of oxygen within and between its three main reservoirs: the atmosphere, the biosphere, and the lithosphere. The main driving factor of the oxygen cycle is photosynthesis, which is responsible for the modern Earth's atmosphere and life as we know it. If all photosynthesis were to cease, the Earth's atmosphere would be devoid of all but trace amounts of oxygen within 5000 years. The oxygen cycle would no longer exist.

***carbon cycle***

The carbon cycle is the biogeochemical cycle by which carbon is exchanged between the biosphere, geosphere, hydrosphere and atmosphere of the Earth. Other bodies may have carbon cycles, but little is known about them. All of these components are reservoirs of carbon. The cycle is usually thought of as four main reservoirs of carbon interconnected by pathways of exchange. The reservoirs are the atmosphere, terrestrial biosphere (usually includes freshwater systems), oceans, and sediments (includes fossil fuels). The annual movements of carbon, the carbon exchanges between reservoirs, occur because of various chemical, physical, geological, and biological processes. The ocean contains the largest pool of carbon near the surface of the Earth, but most of that pool is not involved with rapid exchange with the atmosphere.

The global carbon budget is the balance of the exchanges (incomes and losses) of carbon between the carbon reservoirs or between one specific loop (*e.g.*, atmosphere - biosphere) of the carbon cycle. An examination of the carbon budget of a pool or reservoir can provide information about whether the pool or reservoir is functioning as a source or sink for carbon dioxide.

***phosphorus cycle***

The phosphorus cycle is the biogeochemical cycle that describes the movement of phosphorus through the lithosphere, hydrosphere, and biosphere.

Unlike many other biogeochemicals, the atmosphere does not play a significant role in the movements of phosphorus, because phosphorus and phosphorus-based compounds are usually solids at the typical ranges of temperature and pressure found on Earth.

### *sulphur cycle*

Sulphur is one of the constituents of many proteins and vitamins and hormones. It recycles like other biogeochemical cycles.

The essential steps of the sulphur cycle are:

Mineralization of organic sulphur to the inorganic form, hydrogen sulfide: ($H_2S$).

Oxidation of sulfide and elemental sulphur (S) and related compounds to sulfate ($SO_4^{2-}$).

Reduction of sulfate to sulfide.

Microbial immobilization of the sulphur compounds and subsequent incorporation into the organic form of sulphur.

### *water cycle*

The water cycle — technically known as the hydrologic cycle — is the continuous circulation of water within the Earth's hydrosphere, and is driven by solar radiation. This includes the atmosphere, land, surface water and groundwater.

As water moves through the cycle, it changes state between liquid, solid, and gas phases. Water moves from compartment to compartment, such as from river to ocean, by the physical processes of evaporation, precipitation, infiltration, run-off, and subsurface flow. Movement of water within the water cycle is the subject of the field of hydrology.

### *hydrogen cycle*

Hydrogen is one of the constituents of water. It recycles as in other biogeochemical cycles. It is actively involved with the other cycles like the carbon cycle, nitrogen cycle and sulphur cycle as well.

Anaerobic decomposition of organic substances to carbon dioxide and methane is a collaborative effort involving many different reactions and species of microorganisms. This is also called *interspecies hydrogen transfer.*

## Nutrients for plant growth

Plant nutrition often is confused with fertilization. Plant nutrition refers to a plant's need for and use of basic chemical elements. Fertilization is the term used when these materials are added to the environment around a plant. A lot must happen before a chemical element in a fertilizer can be used by a plant.

Plants need 17 elements for normal growth. Three of them—carbon, hydrogen, and oxygen—are found in air and water. The rest are found in the soil.

Six soil elements are called macronutrients because they are used in relatively large amounts by plants. They are nitrogen, potassium, magnesium, calcium, phosphorus, and sulphur in Table.

**Table. Plant macronutrients.**

| Element | Absorbed as | Leaches from soil/ Mobility in plant | Signs of excess | Signs of deficiency | Notes |
|---|---|---|---|---|---|
| Nitrogen (N) | $NO_3^-$, (nitrate), $NH_4^+$ (ammonium) | Leachable, especially $NO_3^-$. Mobile in plants. | Succulent growth; dark green color; weak, spindly growth; few fruits. May cause brittle growth, especially under high temperatures. | Reduced growth, yellowing (chlorosis). Reds and purples may intensify in some plants. Reduced lateral bud breaks. Symptoms appear first on older growth. | In general, the best $NH_4^+$:$NO_3^-$ ratio is 1:1. Under low sugar conditions (low light), high $NH_4^+$ can cause leaf curl. Uptake is inhibited by high P levels. The N:K ratio is extremely important. Indoors, the best N:K ratio is 1:1 unless light is extremely high. In soils with a high C:N ratio, more N should be supplied. |
| Phosphorus (P) | $H_2PO_4^-$, $HPO_4^-$ (phosphate) | Normally not leachable, but may leach from soil high in bark or peat. Not readily mobile in plants. | Shows up as micronutrient deficiency of Zn, Fe, or Co. | Reduced growth. Color may intensify; browning or purpling of foliage in some plants. Thin stems, reduced lateral bud breaks, loss of lower leaves, reduced flowering. | Rapidly bound (fixed) on soil (P) particles. Under acid conditions, fixed with Fe, Mg, and Al. Under alkaline conditions, fixed with Ca. Important for young plant and seedling growth. High P interferes with micronutrient absorption and N absorption. Used in relatively small amounts when compared to N and K. |
| Potassium (K) | $K^+$ | Can leach in sandy soils. Mobile in plants. | Causes N deficiency in plant and may affect the uptake of other positive ions. | Reduced growth, shortened internodes. Marginal burn or scorch (brown leaf edges), necrotic (dead) spots in leaves. Reduction of lateral bud breaks, tendency to wilt readily. | N:K balance is important. High N:low K favors vegetative growth; low N:high K promotes reproductive growth (flowers, fruit). |

Eight other soil elements are used in much smaller amounts and are called micronutrients or trace elements in Table. They are iron, zinc, molybdenum, manganese, boron, copper, cobalt, and chlorine.

Most of the nutrients a plant needs are dissolved in water and then absorbed by its roots. In fact, 98 per cent are absorbed from the soil-water solution, and only about 2 per cent are actually extracted from soil particles.

**Table. Plant micronutrients.**

| Element | Absorbed as | Signs of excess | Signs of deficiency | Notes |
|---|---|---|---|---|
| Iron (Fe) | $Fe^{++}$, $Fe^{+++}$ | Rare except on flooded soils. Interveinal chlorosis, primarily on young tissue, which eventually may turn white. | Soil high in Ca, Mn, P, or heavy metals (Cu, Zn); high pH; poorly drained soil; oxygen-deficient soil; nematode attack on roots. | Add Fe in the chelate form. The type of chelate needed depends on soil pH. |
| Boron (B) | $BO_3^-$ Borate | Blackening or death of tissue between veins. | Failure to set seed, internal breakdown, death of apical buds. | |
| Zinc (Zn) | $Zn^{++}$ | Shows up as Fe deficiency. Also interferes with Mg absorption. | "Little leaf" (reduction in leaf size), short internodes, distorted or puckered leaf margins, interveinal chlorosis. | |
| Copper (Cu) | $Cu^{++}$, $Cu^{+}$ | Can occur at low pH. Shows up as Fe deficiency. | New growth small, mis-shapen, wilted. | May be found in some peat soils. |
| Manganese (Mn) | $Mn^{++}$ | Reduction in growth, brown spotting on leaves. Shows up as Fe deficiency. | Interveinal chlorosis of leaves followed by brown spots, producing a checkered effect. | Found under acid conditions. |
| Molybdenum (Mo) | $MoO_4^-$ (molybdate) | | Interveinal chlorosis on older or midstem leaves, twisted leaves (whiptail). | |
| Chlorine (Cl) | $Cl^-$ | Salt injury, leaf burn. May increase succulence. | Leaves wilt, then become bronze, then chlorotic, then die; club roots. | |

### *Fertilizers*

Fertilizers are materials containing plant nutrients that are added to the environment around a plant. Generally, they are added to the water or soil, but some can be sprayed on leaves. This method is called foliar fertilization. It should be done carefully with a dilute solution, because a high fertilizer concentration can injure leaf cells. The nutrient, however, does need to pass through the thin layer of wax (cutin) on the leaf surface. Fertilizers are not plant food! Plants produce their own food from water, carbon dioxide, and solar energy through photosynthesis. This food (sugars and carbohydrates) is combined with plant nutrients to produce proteins, enzymes, vitamins, and other elements essential to growth.

### *Nutrient absorption*

Anything that reduces or stops sugar production in leaves can lower nutrient absorption. Thus, if a plant is under stress because of low light or extreme temperatures, nutrient deficiency may develop.

A plant's developmental stage or rate of growth also may affect the amount of nutrients absorbed. Many plants have a rest (dormant) period during part of the year. During this time, few nutrients are absorbed. Plants also may absorb different nutrients as flower buds begin to develop than they do during periods of rapid vegetative growth.

A nutrient is any element or compound necessary for or contributing to an organism's metabolism, growth, or other functioning. Six nutrient groups exist, classifiable as those that provide energy, and as those that otherwise support metabolic processes in the body: Some of them are essential because they cannot be synthesized in the body and must be obtained from a food source.

*Substances that provide energy*

Carbohydrates are compounds made up of sugars. Carbohydrates are classified by their number of sugar units: monosaccharides (such as glucose and fructose), disaccharides (such as sucrose and lactose), oligosaccharides, and polysaccharides (such as starch, glycogen, and cellulose).

Proteins are organic compounds that consists of amino acids joined by peptide bonds. The body does not manufacture certain amino acids (termed essential amino acids); the diet must supply these. In nutrition, proteins are broken down through digestion back into free amino acids.

Fats consist of a glycerin molecule with three fatty acids attached. Fatty acids are unbranched hydrocarbon chains, connected by single bonds alone (saturated fatty acids) or by both single and double bonds (unsaturated fatty acids). Fats are needed to keep cells functioning properly, to insulate body organs against shock, to keep body temperature stable, and to maintain healthy skin and hair. The body does not manufacture certain fatty acids (termed essential fatty acids); the diet must supply these.

Fat has an energy content of 9 kcal/g (~37.7 kJ/g); proteins and carbohydrates 4 kcal/g (~16.7 kJ/g). Ethanol (grain alcohol) has an energy content of 7 kcal/g (~29.3 kJ/g).

*Substances that support metabolism*

Minerals are generally trace elements, salts, or ions such as copper and iron. These minerals are essential to human metabolism.

Vitamins are organic compounds essential to the body. They usually act as coenzymes for various proteins in the body. Water is an essential nutrient and is directly involved in all the chemical reactions of life.[citation needed]

*Nutrition as a science*

Any classification of "nutrients" is likely to be arbitrary given the status of nutrition as a developing science. Researchers are becoming more aware of a wider range of nutrients essential for health.

An organism will metabolise any organic compound to use for its energy content, for structural purposes (growth or replacement of living structures), or for participation in chemical reactions necessary for life. Any particular substance can play more than one role in the body, though researchers lack a good understanding of these roles.

The discovery of the group of nutrients called phytonutrients reinforces the provisional nature of our knowledge. We know little about phytonutrients, organic compounds from plants, which play an essential role in the normal functioning of a body and have complex hormonal effects on health, or play an active role in the amelioration of disease. They do not fit readily into the scheme of the traditional nutrition categories.

*Nutrients and the environment*

The band of a green alga (*Enteromorpha*) along this shore indicates that there is a nearby source of nutrients (probably nitrates or ammonia from a small estuary).

While in essence true to the definition above, the term nutrients has a more limited meaning within the specialised fields of water quality and water pollution, referring specifically to plant fertilizers. In this context, certain mineral compounds can have an adverse impact on water quality because of their ability to promote plant and algae growth. An excessive growth of aquatic plants can clog waterways, and over-stimulation of algae and microbes leads to an ecological process called eutrophication.

A surprisingly small number of elements provide interest or concern in this context: really just nitrogen and phosphorus in most aquatic systems. Mineral compounds involved are ammonia, nitrites, nitrates, and orthophosphates. Organic compounds also may contribute, in as much as they also contain nitrogen and phosphorus. The reason only a few chemicals are of concern has to do with the fact that plants are made up mostly of compounds of carbon (C), hydrogen (H), oxygen (O), nitrogen (N), and phosphorus (P), and lesser amounts of sulphur (S), potassium (K), magnesium (Mg), and calcium (Ca). These elements constitute the macronutrients. Many other elements, though necessary for growth, classify as micronutrients due to the very small quantities required.

Plants obtain carbon, hydrogen, and oxygen (elements most needed for growth) from the air and water, where all three elements occur in great abundance as water and as carbon dioxide. Nutrients having greatest potential to influence plant growth in aquatic environments is those elements needed for plant growth in proportionately large amounts (that is, macronutrients) but likely to become limiting—that is, present in amounts that could be depleted by continued growth. Once used up, further growth will not be possible. Of the nine macronutrients, nitrogen and phosphorus are most likely to become limiting. The others always remain present in great abundance (C, H, O) or usually in amounts that exceed the requirements of aquatic plants or algae.

Farmers apply fertilizer nutrients in the form of nitrogen, phosphorus, and potassium (N, P, and K with perhaps micronutrients) to prevent these elements from becoming limiting in the soil. These elements become concentrated in wastewaters from animal pens and septic or sewage systems. And these elements (especially N and P) in run-off or wastewater discharges reaching streams, lakes, or seas will promote aquatic plant growth. Abundant plant growth itself gives cause for concern in assessing water quality. The most abundant "plants" in most aquatic environments are algae. When essential nutrients are plentiful, algae multiply. If these algae are microscopic phytoplankton, their growth increases the turbidity of the water. The water then becomes cloudy, coloured a shade of green, yellow, or brown (sometimes red). A super abundance of algae, or of higher plants, in an aquatic system can signal excessive inputs of nutrients.

## The pathway of nitrogen

### *Role of nitrogen in the biosphere*

The growth of all organisms depends on the availability of mineral nutrients, and none is more important than nitrogen, which is required in large amounts as an essential component of proteins, nucleic acids and other cellular constituents.

There is an abundant supply of nitrogen in the earth's atmosphere - nearly 79 per cent in the form of N2 gas. However, N2 is unavailable for use by most organisms because there is a triple bond between the two nitrogen atoms, making the molecule almost inert. In order for nitrogen to be used for growth it must be "fixed" (combined) in the form of ammonium (NH4) or nitrate (NO3) ions. The weathering of rocks releases these ions so slowly that it has a neglible effect on the availability of fixed nitrogen. So, nitrogen is often the limiting factor for growth and biomass production in all environments where there is suitable climate and availability of water to support life.

Microorganisms have a central role in almost all aspects of nitrogen availability and thus for life support on earth:

Some bacteria can convert N2 into ammonia by the process termed nitrogen fixation; these bacteria are either free-living or form symbiotic associations with plants or other organisms (*e.g.* termites, protozoa) other bacteria bring about transformations of ammonia to nitrate, and of nitrate to N2 or other nitrogen gases many bacteria and fungi degrade organic matter, releasing fixed nitrogen for reuse by other organisms.

All these processes contribute to the nitrogen cycle.

We shall deal first with the process of nitrogen fixation and the nitrogen-fixing organisms, then consider the microbial processes involved in the cycling of nitrogen in the biosphere.

*Nitrogen fixation*

A relatively small amount of ammonia is produced by lightning. Some ammonia also is produced industrially by the Haber-Bosch process, using an iron-based catalyst, very high pressures and fairly high temperature. But the major conversion of N2 into ammonia, and thence into proteins, is achieved by microorganisms in the process called nitrogen fixation (or dinitrogen fixation).

The table below shows some estimates of the amount of nitrogen fixed on a global scale. The total biological nitrogen fixation is estimated to be twice as much as the total nitrogen fixation by non-biological processes.

**Type of fixation N2 fixed (1012 g per year, or 106 metric tons per year)**

| **Type of fixation** | **$N_2$ fixed ($10^{12}$ g per year, or $10^6$ metric tons per year)** |
|---|---|
| Non-biological | Â> |
| Non-biological | Â |
| Industrial | about 50 |
| Combustion | about 20 |
| Lightning | about 10 |
| Total | about 80 |
| Â | Â |
| Biological | Â |
| Agricultural land | about 90 |
| Forest and non-agricultural land | about 50 |
| Sea | about 35 |
| Total | about 175 |

To illustrate the importance of biological nitrogen fixation, the image below shows part of the Lower Sonoran desert in Arizona. Every plant in this scene depends ultimately on biological nitrogen fixation. Both free-living cyanobacteria and the cyanobacterial associates of lichens initially contributed nitrogen to the soil by forming a cryptobiotic crust. Now numerous leguminous plants occur in this desert, with nitrogen-fixing Rhizobium in their root nodules. Examples are the green-stemmed brush-like trees at the right and left of the image (Parkinsonia species, common name "paloverde"), and several acacias and mesquites.

*Mechanism of biological nitrogen fixation*

Biological nitrogen fixation can be represented by the following equation, in which two moles of ammonia are produced from one mole of nitrogen gas, at the expense of 16 moles of ATP and a supply of electrons and protons (hydrogen ions):

$N_2 + 8H+ + 8e- + 16\ ATP = 2NH_3 + H_2 + 16ADP + 16\ Pi$

This reaction is performed exclusively by prokaryotes (the bacteria and related organisms), using an enzyme complex termed nitrogenase. This enzyme consists of two proteins - an iron protein and a molybdenum-iron protein, as shown below.

The reactions occur while N2 is bound to the nitrogenase enzyme complex. The Fe protein is first reduced by electrons donated by ferredoxin. Then the reduced Fe protein binds ATP and reduces the molybdenum-iron protein, which donates electrons to $N_2$, producing HN=NH. In two further cycles of this process (each requiring electrons donated by ferredoxin) HN=NH is reduced to $H_2N\text{-}NH_2$, and this in turn is reduced to $2NH_3$.

Depending on the type of microorganism, the reduced ferredoxin which supplies electrons for this process is generated by photosynthesis, respiration or fermentation.

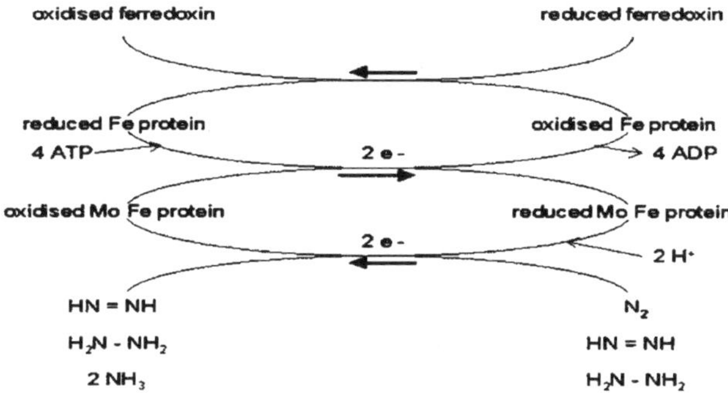


There is a remarkable degree of functional conservation between the nitrogenase proteins of all nitrogen-fixing bacteria. The Fe protein and the Mo-Fe protein have been isolated from many of these bacteria, and nitrogen fixation can be shown to occur in cell-free systems in a laboratory when the Fe protein of one species is mixed with the Mo-Fe protein of another bacterium, even if the species are very distantly related.

*The nitrogen-fixing organisms*

All the nitrogen-fixing organisms are prokaryotes (bacteria). Some of them live independently of other organisms - the so-called free-living nitrogen-fixing bacteria. Others live in intimate symbiotic associations with plants or with other organisms (*e.g.* protozoa). Examples are shown in the table below.

A point of special interest is that the nitrogenase enzyme complex is highly sensitive to oxygen. It is inactivated when exposed to oxygen, because this reacts with the iron component of the proteins. Although this is not a problem for anaerobic bacteria, it could be a major problem for the aerobic species such as cyanobacteria (which generate oxygen during photosynthesis) and the free-living aerobic bacteria of soils, such as Azotobacter and Beijerinckia. These organisms have various methods to overcome the problem. For example, Azotobacter species have the highest known rate of respiratory metabolism of any organism, so they might protect the enzyme by maintaining a very low level of oxygen in their cells. Azotobacter species also produce copious amounts of extracellular polysaccharide (as do Rhizobium species in culture - ). By maintaining water within the polysaccharide slime layer, these bacteria can limit the diffusion rate of oxygen to the cells. In the symbiotic nitrogen-fixing organisms such as Rhizobium, the root nodules can contain oxygen-scavenging molecules such as leghaemoglobin, which shows as a pink colour when the active nitrogen-fixing nodules of legume roots are cut open. Leghaemoglobin may regulate the supply of oxygen to the nodule tissues in the same way as haemoglobin regulates the supply of oxygen to mammalian tissues. Some of the cyanobacteria have yet another mechanism for protecting nitrogenase: nitrogen fixation occurs in special cells (heterocysts) which possess only photosystem I (used to generate ATP by light-mediated reactions) whereas the other cells have both photosystem I and photosystem II (which generates oxygen when light energy is used to split water to supply H2 for synthesis of organic compounds).

*Symbiotic nitrogen fixation*

The most familiar examples of nitrogen-fixing symbioses are the root nodules of legumes (peas, beans, clover, etc.).

Part of a clover root system bearing naturally occurring nodules of Rhizobium. Each nodule is about 2-3 mm long.

Clover root nodules at higher magnification, showing two partly crushed nodules (arrowheads) with pink-coloured contents. This colour is caused by the presence of the pigment leghaemoglobin - a unique metabolite of this type of symbiosis. Leghaemoglobin is found only in the nodules and is not produced by either the bacterium or the plant when grown alone.

In these leguminous associations the bacteria usually are Rhizobium species, but the root nodules of soybeans, chickpea and some other legumes are formed by small-celled rhizobia termed Bradyrhizobium. Nodules on some tropical leguminous plants are formed by yet other genera. In all cases the bacteria "invade" the plant and cause the formation of a nodule by inducing localised proliferation of the plant host cells. Yet the bacteria always remain separated from the host cytoplasm by being enclosed in a membrane - a necessary feature in symbioses.

Part of a crushed root nodule of a pea plant, showing four root cells containing colonies of Rhizobium. The nuclei (n) of two root cells are shown; cw indicates the cell wall that separates two plant cells. Although it cannot be seen clearly in this image, the bacteria occur in clusters which are enclosed in membranes, separating them from the cytoplasm of the plant cells.

In nodules where nitrogen-fixation is occurring, the plant tissues contain the oxygen-scavenging molecule, leghaemoglobin (serving the same function as the oxygen-carrying haemoglobin in blood). The function of this molecule in nodules is to reduce the amount of free oxygen, and thereby to protect the nitrogen-fixing enzyme nitrogenase, which is irreversibly inactivated by oxygen.

Some excellent images and discussion of these leguminous associations can be found at:

Frankia is a genus of the bacterial group termed actinomycetes - filamentous bacteria that are noted for their production of air-borne spores. Included in this group are the common soil-dwelling Streptomyces species which produce many of the antibiotics used in medicine.

Frankia species are slow-growing in culture, and require specialised media, suggesting that they are specialised symbionts. They form nitrogen-fixing root nodules (sometimes called actinorhizae) with several woody plants of different families, such as alder (Alnus species), sea buckthorn (Hippophae rhamnoides, which is common in sand-dune environments) and Casuarina (a Mediterranean tree genus).

Alder and the other woody hosts of Frankia are typical pioneer species that invade nutrient-poor soils. These plants probably benefit from the nitrogen-fixing association, while supplying the bacterial symbiont with photosynthetic products.

The photosynthetic cyanobacteria often live as free-living organisms in pioneer habitats such as desert soils or as symbionts with lichens in other pioneer habitats. They also form symbiotic associations with other organisms such as the water fern Azolla, and cycads.The association with Azolla, where cyanobacteria (Anabaena azollae) are harboured in the leaves, has sometimes been shown to be important for nitrogen inputs in rice paddies, especially if the fern is allowed to grow and then ploughed into the soil to release nitrogen before the rice crop is sown. A symbiotic association of cyanobacteria with cycads is shown below. The first image shows a pot-grown plant. The second image shows a close-up of the soil surface in this pot. Short, club-shaped, branching roots have grown into the aerial environment. These aerial roots contain a nitrogen-fixing cyanobacterial symbiont.

In addition to these intimate and specialised symbiotic associations, there are several free-living nitrogen-fixing bacteria that grow in close association with plants. For example, Azospirillum species have been shown to fix nitrogen when growing in the root zone (rhizosphere) or tropical grasses, and even of maize plants in field conditions. Similarly, Azotobacter species can fix nitrogen in the rhizosphere of several plants. In both cases the bacteria grow at the expense of sugars and other nutrients that leak from the roots. However, these bacteria can make only a small contribution to the nitrogen nutrition of the plant, because nitrogen-fixation is an energy-expensive process, and large amounts of organic nutrients are not continuously available to microbes in the rhizosphere.

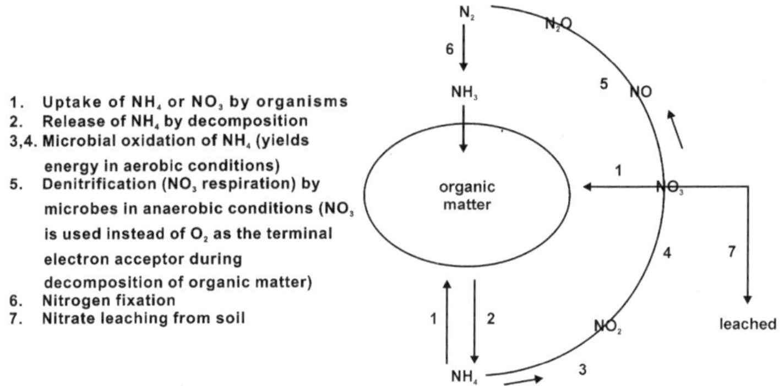


This limitation may not apply to the bacteria that live in root nodules or other intimate symbiotic associations with plants. It has been estimated that nitrogen fixation in the nodules of clover roots or other leguminous plants may consume as much as 20 per cent of the total photosynthate.

*The nitrogen cycle*

At any one time a large proportion of the total fixed nitrogen will be locked up in the biomass or in the dead remains of organisms. So, the only nitrogen available to support new growth will be that which is supplied by nitrogen fixation from the atmosphere or by the release of ammonium or simple organic nitrogen compounds through the decomposition of organic matter.

*Nitrification*

The term nitrification refers to the conversion of ammonium to nitrate (pathway 3-4). This is brought about by the nitrifying bacteria, which are specialised to gain their energy by oxidising ammonium, while using CO2 as their source of carbon to synthesise organic compounds. Organisms of this sort are termed chemoautotrophs - they gain their energy by chemical oxidations (chemo-) and they are autotrophs (self-feeders) because they do not depend on pre-formed organic matter. In principle the oxidation of ammonium by these bacteria is no different from the way in which humans gain energy by oxidising sugars. Their use of $CO_2$ to produce organic matter is no different in principle from the behaviour of plants.

The nitrifying bacteria are found in most soils and waters of moderate pH, but are not active in highly acidic soils. They almost always are found as mixed-species communities (termed consortia) because some of them - *e.g.* Nitrosomonas species - are specialised to convert ammonium to nitrite ($NO_2$-) while others - *e.g.* Nitrobacter species - convert nitrite to nitrate ($NO_3$-). In fact, the accumulation of nitrite inhibits Nitrosomonas, so it depends on Nitrobacter to convert this to nitrate, whereas Nitrobacter depends on Nitrosomonas to generate nitrite.

The nitrifying bacteria have some important environmental consequences, because they are so common that most of the ammonium in oxygenated soil or natural waters is readily converted to nitrate. Most plants and microorganisms can take up either nitrate or ammonium. However, process of nitrification has some undesirable consequences. The ammonium ion ($NH_4$+) has a positive charge and so is readily adsorbed onto the negatively charged clay colloids and soil organic matter, preventing it from being washed out of the soil by rainfall. In contrast, the negatively charged nitrate ion is not held on soil particles and so can be washed down the soil profile - the process termed leaching. In this way, valuable nitrogen can be lost from the soil, reducing the soil fertility. The nitrates can then accumulate in groundwater, and ultimately in drinking water. There are strict regulations governing the amount of nitrate that can be present in drinking water, because nitrates can be reduced to highly reactive nitrites by microorganisms in the anaerobic conditions of the gut. Nitrites are absorbed from the gut and bind to haemoglobin, reducing its oxygen-carrying capacity. In young babies this can lead to respiratory distress - the condition known as "blue baby syndrome". Nitrite in the gut also can react with amino compounds, forming highly carcinogenic nitrosamines.

*Denitrification*

Denitrification refers to the process in which nitrate is converted to gaseous compounds (nitric oxide, nitrous oxide and $N_2$) by microorganisms. The sequence usually involves the production of nitrite ($NO_2$-) as an intermediate step.Several types of bacteria perform this conversion when growing on organic matter in anaerobic conditions. Because of the lack of oxygen for normal aerobic respiration, they use nitrate in place of oxygen as the terminal electron acceptor. This is termed anaerobic respiration and can be illustrated as follows:

In aerobic respiration (as in humans), organic molecules are oxidised to obtain energy, while oxygen is reduced to water:

$$C_6H_{12}O_6 + 6\ O_2 = 6\ CO_2 + 6\ H_2O + \text{energy}$$

In the absence of oxygen, any reducible substance such as nitrate ($NO_3$-) could serve the same role and be reduced to nitrite, nitric oxide, nitrous oxide or $N_2$.

Thus, the conditions in which we find denitrifying organisms are characterised by (1) a supply of oxidisable organic matter, and (2) absencc of oxygen but availability of reducible nitrogen sources. A mixture of gaseous nitrogen products is often produced because of the stepwise use of nitrate, nitrite, nitric oxide and nitrous oxide as electron acceptors in anaerobic respiration. The common denitrifying bacteria include several species of Pseudomonas, Alkaligenes and Bacillus. Their activities result in substantial losses of nitrogen into the atmosphere, roughly balancing the amount of nitrogen fixation that occurs each year.

## Phosphorus and sulphur

### *Phosphorus*

Phosphorus, (from the Greek language *phôs* meaning "light", and *phoros* meaning "bearer"), is the chemical element in the periodic table that has the symbol P and atomic number 15. A multivalent non-metal of the nitrogen group, phosphorus is commonly found in inorganic phosphate rocks and in all living cells. Due to its high reactivity, it is never found as a free element in nature. It emits a faint glow upon exposure to oxygen, occurs in several allotropic forms, and is an essential element for living organisms. The most important commercial use of phosphorus is in the production of fertilizers. It is also widely used in explosives, nerve agents, friction matches, fireworks, pesticides, toothpaste, and detergents.

Phosphorus (P), the 15th element on the periodic table with an atomic weight of 30.974, is an essential nutrient for all life forms. Phosphorus plays a role in deoxyribonucleic acid (DNA), ribonucleic acid (RNA), adenosine diphosphate (ADP), and adenosine triphosphate (ATP). Phosphorus is required for these necessary components of life to occur.

Phosphate is usually not readily available for uptake in soils. Phosphate is only freely soluble in acid solutions and under reducing conditions. In the soil it is rapidly immobilized as calcium or iron phosphates. Most of the phosphorus in soils is adsorbed to soil particles or incorporated into organic matter.

Phosphorus in freshwater and marine systems exists in either a particulate phase or a dissolved phase. Particulate matter includes living and dead plankton, precipitates of phosphorus, phosphorus adsorbed to particulates, and amorphous phosphorus. The dissolved phase includes inorganic phosphorus (generally in the soluble orthophosphate form), organic phosphorus excreted by organisms, and macromolecular colloidal phosphorus.

The organic and inorganic particulate and soluble forms of phosphorus undergo continuous transformations. The dissolved phosphorus (usually as orthophosphate) is assimilated by phytoplankton and altered to organic phosphorus. The phytoplankton are then ingested by detritivores or zooplankton. Over half of the organic phosphorus taken up by zooplankton is excreted as inorganic P. Continuing the cycle, the inorganic P is rapidly assimilated by phytoplankton.

Lakes and reservoir sediments serve as phosphorus sinks. Phosphorus-containing particles settle to the substrate and are rapidly covered by sediment. Continuous accumulation of sediment will leave some phosphorus too deep within the substrate to be reintroduced to the water column. Thus, some phosphorus is removed permanently from biocirculation.

*Health Effects*

*Phosphate*: Phosphate itself does not have notable adverse health effects. However, phosphate levels greater than 1.0 may interfere with coagulation in water treatment plants. As a result, organic particles that harbour microorganisms may not be completely removed before distribution.

*Environmental Effects*

The growth of macrophytes and phytoplankton is stimulated principally by nutrients such as phosphorus and nitrogen. Nutrient-stimulated primary production is of most concern in lakes and estuaries, because primary production in flowing water is thought to be controlled by physical factors, such as light penetration, timing of flow, and type of substrate available, instead of by nutrients.

Freshwater system impacts: Generally, phosphorus (as orthophosphate) is the limiting nutrient in freshwater aquatic systems. That is, if all phosphorus is used, plant growth will cease, no matter how much nitrogen is available. The natural background levels of total phosphorus are generally less than 0.03 mg/l. The natural levels of orthophosphate usually range from 0.005 to 0.05 mg/l Many bodies of freshwater are currently experiencing influxes of phosphorus and nitrogen from outside sources. The increasing concentration of available phosphorus allows plants to assimilate more nitrogen before the phosphorus is depleted. Thus, if sufficient phosphorus is available, elevated concentrations of nitrates will lead to algal blooms. Although levels of 0.08 to 0.10 mg/l orthophosphate may trigger periodic blooms, long-term eutrophication will usually be prevented if total phosphorus levels and orthophosphate levels are below 0.5 mg/l and 0.05 mg/l, respectively.

Estuarine system impacts: In contrast to freshwater, nitrogen is generally the primary limiting nutrient in the seaward portions of estuarine systems. Here, nitrogen levels control the rate of primary production. If the system is supplied with high levels of nitrogen, algal blooms will occur. Systems may be phosphorus limited, however, or become so when nitrogen concentrations are high and $N:P>16:1$.

Freshwater and estuarine systems: Nutrient-induced production of aquatic plants in both freshwater and estuaries has several detrimental consequences:

Extensive growth of rooted aquatic macrophytes will interfere with navigation, aeration, and channel capacity.

Dead macrophytes and phytoplankton settle to the bottom of a water body, stimulating microbial breakdown processes that require oxygen. Eventually, oxygen will be depleted.

*Sources*

*Non-point sources:*

Natural: Phosphate deposits and phosphate-rich rocks release phosphorus during weathering, erosion, and leaching. Phosphorus may be released from lake and reservoir bottom sediments during seasonal overturns.

Anthropogenic: The primary anthropogenic non-point sources of phosphorus include run-off from 1) land areas being mined for phosphate deposits, 2) agricultural areas, and 3) urban/residential areas. Because phosphorus has a strong affinity for soil, little dissolved phosphorus will be transported in run-off. Instead, the eroded sediments from mining and agricultural areas carry the adsorbed phosphorus to the water body. An additional source is the overboard discharge of phosphorus-containing sewage by boats.

*Point sources:* Sewage treatment plants provide most of the available phosphorus to surface water bodies. A normal adult excretes 1.3 - 1.5 g of phosphorus per day. Additional phosphorus originates from the use of industrial products, such as toothpaste, detergents, pharmaceuticals, and food-treating compounds. Primary treatment removes only 10 per cent of the phosphorus in the waste stream; secondary treatment removes only 30 per cent. The remainder is discharged to the water body (Smith, 1990). Tertiary treatment is required to remove additional phosphorus from the water. The amount of additional phosphorus that can be removed varies with the success of the treatment technologies used. Available technologies include biological removal and chemical precipitation.

*Mode of Transport*

Phosphates are primarily discharged directly into the water body by sewage treatment plants. Phosphorus that is adsorped to sediment particles may be transported in overland flow.

*Analytical techniques:*

A. Total Phosphorus and Orthophosphate: Analysis involves two procedural steps: 1) conversion of the phosphorus form into dissolved orthophosphate by a digestion method, and 2) colourimetric evaluation of the dissolved orthophosphate concentration.

*Step 1: Digestion methods*

Perchloric Acid Digestion: Recommended only for extremely difficult-to-analyze samples, such as sediments.

Nitric Acid-Sulphuric Acid Method Recommended for most samples.

Persulfate Oxidation Method This simple method should be cross-checked with one or more thorough techniques and adopted if results are identical.

*Step 2: Colorimetric methods*

Ascorbic Acid Method: Ammonium molybdate and potassium antimonyl tartrate react with orthophosphate to form a heteropoly acid that is reduced to molybdenum blue by ascorbic acid.

*Detection limits:* Ranges change with light path used.

| Range (mg/l P) | Path (cm) |
|---|---|
| 0.3 - 2.0 | 0.5 |
| 0.15 - 1.3 | 1.0 |
| 0.01 - 0.25 | 5.0 |

*Interferences:* Arsenates react with the molybdate to form a similar blue colour. Nitrite and hexavalent chromium interfere to yield results 3 per cent less than actual at 1 mg/l and 10 per cent to 15 per cent less than actual at 10 mg/l.

Automated Ascorbic Acid Reduction Method: Ammonium molybdate and potassium antimonyl tartrate react with orthophosphate in an acid medium to form an antimony- phosphomolybdate complex that forms a blue colour suitable for photometric measurements when reduced by ascorbic acid.

*Detection limits:* 0.001 to 10.0 mg/l P when photometric measurements are performed at 650 to 600 in a 15mm tubular flow cell, or 880 nm in a 50mm tubular flow cell.

*Interferences:* >50 mg/l Fe(3+), 10 mg/l Cu, and 10 mg/l SiO2. Turbidity, colour may interfere. Arsenate provides a positive interference.

*Notable characteristics*

Common phosphorus forms a waxy white solid that has a characteristic, disagreeable smell similar to that of garlic. Pure forms of the element are colorless and transparent. This non-metal is not soluble in water, but *is* soluble in carbon disulfide. Pure phosphorus ignites spontaneously in air, burning to produce phosphorus pentoxide.

*Forms*

Phosphorus is known to exist in three allotropic forms: white, red, and black (other allotropic forms are theoretically possible, but are not known). The most common are red and white phosphorus, both of which consist of networks of tetrahedrally arranged groups of four phosphorus atoms. The tetrahedra of white phosphorus form separate groups; the tetrahedra of red phosphorus are linked into chains. White phosphorus burns on contact with air and on exposure to heat or light. Phosphorus also exists in kinetically and thermodynamically favored forms. They are separated by a transition temperature of -3.8 °C. One is known as the "alpha" form, the other "beta". Red phosphorus is comparatively stable and sublimates at a vapor pressure of 1 atm at 170 °C but burns from impact or frictional heating. A black phosphorus allotrope exists which has a structure similar to graphite – the atoms are arranged in hexagonal sheet layers and will conduct electricity.

*Glow*

The glow from phosphorus was the attraction of its discovery around 1669, but the mechanism for that glow was not fully described until 1974.

In 1974 the glow was explained by R. J. van Zee and A. U. Khan. A reaction with oxygen takes place at the surface of the solid (or liquid) phosphorus, forming short-lived molecules HPO and $P_2O_2$ and they both emit visible light. The reaction is slow and only very little of the intermediates is required to produce the luminescence, hence the extended time the glow continues in a stoppered jar.

Although the term phosphorescence is derived from phosphorus, the reaction is properly called luminescence (glowing by its own reaction, in this case chemoluminescence), not phosphorescence (re-emitting light that previously fell on it).

*Applications*

Concentrated phosphoric acids, which can consist of 70 per cent to 75 per cent $P_2O_5$ are very important to agriculture and farm production in the form of fertilizers. Global demand for fertilizers led to large increases in phosphate ($PO_4^{3-}$) production in the second half of the 20th century. Other uses;

Phosphates are utilized in the making of special glasses that are used for sodium lamps.

Phosphorus is widely used to make organophosphorus compounds, through the intermediates phosphorus chlorides and the two phosphorus sulfides: phosphorus pentasulfide, and phosphorus sesquisulfide. Organophosphorus compounds have many applications, including in plasticizers, flame retardants, pesticides, extraction agents, and water treatment.

Phosphorus sesquisulfide is used in heads of strike-anywhere matches.

This element is also an important component in steel production, in the making of phosphor bronze, and in many other related products.

White phosphorus is used in military applications as incendiary bombs, for smoke-screening as smoke pots and smoke bombs, and in tracer ammunition.

Red phosphorus is essential for manufacturing matchbook strikers, flares, and, most notoriously, methamphetamine.

In trace amounts, phosphorus is used as a dopant for N-type semiconductors. $^{32}P$ and $^{33}P$ are used as radioactive tracers in biochemical laboratories.

*Biological role*

Phosphorus is a key element in all known forms of life. Inorganic phosphorus in the form of the phosphate $PO_4^{3-}$ plays a major role in biological molecules such as DNA and RNA where it forms part of the structural backbone of these molecules. Living cells also utilize phosphate to transport cellular energy via adenosine triphosphate (ATP). Nearly every cellular process that uses energy gets it in the form of ATP. ATP is also important for phosphorylation, a key regulatory event in cells. Phospholipids are the main structural components of all cellular membranes. Calcium phosphate salts are used by animals to stiffen their bones. An average person contains a little less than 1 kg of phosphorus, about three quarters of which is present in bones and teeth in the form of apatite. A well-fed adult in the industrialized world consumes and excretes about 1-3 g of phosphorus per day in the form of phosphate. Phosphorus is an essential mineral macronutrient, which is studied extensively in soil conservation in order to understand plant uptake from soil systems.

In ecological terms, phosphorus is often a limiting nutrient in many environments, *i.e.* the availability of phosphorus governs the rate of growth of many organisms. In ecosystems an excess of phosphorus can be problematic, especially in aquatic systems.

*History*

Phosphorus (Greek *phosphoros* was the ancient name for the planet Venus) was discovered by German alchemist Hennig Brand in 1669 through a preparation from urine. Working in Hamburg, Brand attempted to distill salts by evaporating urine, and in the process produced a white material that glowed in the dark and burned brilliantly. Since that time, phosphorescence has been used to describe substances that shine in the dark without burning.

Phosphorus was first made commercially, for the match industry, in the 19th century, by distilling off phosphorus vapour from precipitated phosphates heated in a retort. The precipitated phosphates made from ground-up bones, that had been de-greased and treated with strong acids. This process became obsolete in the late 1890s when the Electric arc furnace was adapted to reduce phosphate rock.

Early matches used white phosphorus in their composition, which was dangerous due to its toxicity. Murders, suicides and accidental poisonings resulted from its use. The electric furnace method allowed production to increase to the point phosphorus could be used in weapons of war. In World WarI it was used in incendiaries, smoke screens and tracer bullets.

Today phosphorus production is larger than ever, used as a precursor for various chemicals, in particular the pesticide glyphosate sold under the brand name Roundup. Production takes place at large facilities and is transported heated to liquid form.

*Occurrence*

Due to its reactivity to air and many other oxygen containing substances, phosphorus is not found free in nature but it is widely distributed in many different minerals. Phosphate rock, which is partially made of apatite (an impure tri-calcium phosphate mineral), is an important commercial source of this element. Large deposits of apatite are located in China, Russia, Morocco, Florida, Idaho, Tennessee, Utah, and elsewhere. There are however concerns over how long these phosphorus deposits will last. The United States will deplete its deposits around 2035. China and Morocco have the largest known deposits today, but they too will eventually be depleted. During that depletion there could be a serious problem for the world's food production since phosphorus is such an essential ingredient in fertilizers.

The white allotrope can be produced using several different methods. In one process, tri-calcium phosphate, which is derived from phosphate rock, is heated in an electric or fuel-fired furnace in the presence of carbon and silica. Elemental phosphorus is then liberated as a vapor and can be collected under phosphoric acid.

*Precautions*

Organic compounds of phosphorus form a wide class of materials, some of which are extremely toxic. Fluorophosphate esters are among the most potent neurotoxins known. A wide range of organophosphorus compounds are used for their toxicity to certain organisms as pesticides (herbicides, insecticides, fungicides etc) and weaponized as nerve agents. Most inorganic phosphates are relatively non-toxic and essential nutrients.

The allotrope white phosphorus should be kept under water at all times as it presents a significant fire hazard due to its extreme reactivity to atmospheric oxygen, and it should only be manipulated with forceps since contact with skin can cause severe burns. Chronic white phosphorus poisoning of unprotected workers leads to necrosis of the jaw called "phossy-jaw". Ingestion of white phosphorus may cause a medical condition known as "Smoking Stool Syndrome".

When the white form is exposed to sunlight or when it is heated in its own vapor to 250 °C, it is transmuted to the red form, which does not phosphoresce in air. The red allotrope does not spontaneously ignite in air and is not as dangerous as the white form. Nevertheless, it should be handled with care because it does revert to white phosphorus in some temperature ranges and it also emits highly toxic fumes that consist of phosphorus oxides when it is heated.

Upon exposure to elemental phosphorus, in the past it was suggested to wash the affected area with 2 per cent copper sulfate solution to form harmless compounds that can be washed away. The manual suggests instead "a bicarbonate solution to neutralize phosphoric acid, which will then allow removal of visible WP. Particles often can be located by their emission of smoke when air strikes them, or by their phosphorescence in the dark. In dark surroundings, fragments are seen as luminescent spots." Then, "Promptly debride the burn if the patient's condition will permit removal of bits of WP which might be absorbed later and possibly produce systemic poisoning. DO NOT apply oily-based ointments until it is certain that all WP has been removed. Following complete removal of the particles, treat the lesions as thermal burns." As white phosphorus readily mixes with oils, any oily substances or ointments are disrecommended until the area is thoroughly cleaned and all white phosphorus removed.

*Isotopes*

Radioactive isotopes of phosphorus include:

$^{32}P$; a beta-emitter (1.71 MeV) with a half-life of 14.3 days which is used routinely in life-science laboratories, primarily to produce radiolabeled DNA and RNA probes, *e.g.* for use in Northern blots or Southern blots. Because the high energy beta particles produced penetrate skin and corneas, and because any $^{32}P$ ingested, inhaled, or absorbed is readily incorporated into bone and nucleic acids, OSHA requires that a lab coat, disposable gloves, and safety glasses or goggles be worn when working with $^{32}P$, and that working directly over an open container be avoided in order to protect the eyes. Monitoring personal, clothing, and surface contamination is also required. In addition, due to the high energy of the beta particles, shielding this radiation with the normally used dense materials (*e.g.* lead), gives rise to secondary emission of X-rays via a process known as Bremsstrahlung, meaning braking radiation. Therefore shielding must be accomplished with low density materials, *e.g.* Plexiglas, acrylic, Lucite, plastic, wood, or water.

$^{33}P$; a beta-emitter (0.25 MeV) with a half-life of 25.4 days. It is used in life-science laboratories in applications in which lower energy beta emissions are advantageous such as DNA sequencing.

*Compounds*

Ammonium phosphate ($(NH_4)_3PO_4$)
Phosphine (Phosphorus Trihydride $PH_3$)
Calcium phosphate ($Ca_3(PO_4)_2$)
Calcium dihydrogen phosphate ($Ca(H_2PO_4)_2$)
Iron(III) phosphate ($FePO_4$)
Iron(II) phosphate ($Fe_3(PO_4)_2$)
Hypophosphorous acid ($H_3PO_2$)

***Nature of Phosphorus in Soils***

Phosphorus (P) is an essential element classified as a macronutrient because of the relatively large amounts of P required by plants. Phosphorus is one of the three nutrients generally added to soils in fertilizers. One of the main roles of P in living organisms is in the transfer of energy. Organic compounds that contain P are used to transfer energy from one reaction to drive another reaction within cells. Adequate P availability for plants stimulates early plant growth and hastens maturity. Although P is essential for plant growth, mismanagement of soil P can pose a threat to water quality. The concentration of P is usually sufficiently low in fresh water so that algae growth is limited. When lakes and rivers are polluted with P, excessive growth of algae often results. High levels of algae reduce water clarity and can lead to decreases in available dissolved oxygen as the algae decays, conditions that can be very detrimental to game fish populations.

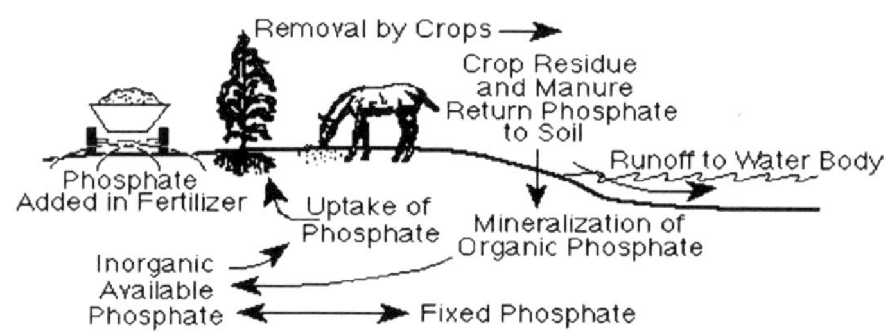


**Fig.** The phosphorus cycle.

Phosphate is taken up by plants from soils, utilized by animals that consume plants, and returned to soils as organic residues decay in soils. Much of the phosphate used by living organisms becomes incorporated into organic compounds.

When plant materials are returned to the soil, this organic phosphate will slowly be released as inorganic phosphate or be incorporated into more stable organic materials and become part of the soil organic matter. The release of inorganic phosphate from organic phosphates is called mineralization and is caused by microorganisms breaking down organic compounds. The activity of microorganisms is highly influenced by soil temperature and soil moisture. The process is most rapid when soils are warm and moist but well drained. Phosphate can potentially be lost through soil erosion and to a lesser extent to water running over or through the soil.

Many phosphate compounds are not very soluble in water; therefore, most of the phosphate in natural systems exists in solid form. However, soil water and surface water (rivers and lakes) usually contain relatively low concentrations of dissolved (or soluble) phosphorus. Depending on the types of minerals in the area, bodies of water usually contain about 10 ppb or more of dissolved P as orthophosphate. Water bodies may also contain organic P and phosphate attached to small particles of sediment. Total phosphorus in water is all of the phosphorus in solution regardless of its form and is often the form reported in water quality studies. Algal available or bioavailable phosphorus is P that is estimated to be available to organisms like algae that are present in a lake or river. This is usually estimated by a chemical test which is designed to measure the dissolved P and the particulate P that are easily available. This is a measure of the P that is of immediate concern to water quality.

*Forms of Phosphorus in Soils*

In soils P may exist in many different forms. In practical terms, however, P in soils can be thought of existing in 3 "pools":

Solution P
Active P
Fixed P

The solution P pool is very small and will usually contain only a fraction of a pound of P per acre. The solution P will usually be in the orthophosphate form, but small amounts of organic P may exist as well. Plants will only take up P in the orthophosphate form. The solution P pool is important because it is the pool from which plants take up P and is the only pool that has any measurable mobility. Most of the P taken up by a crop during a growing season will probably have moved only an inch or less through the soil to the roots. A growing crop would quickly deplete the P in the soluble P pool if the pool was not being continuously replenished.

The active P pool is P in the solid phase which is relatively easily released to the soil solution, the water surrounding soil particles. As plants take up phosphate, the concentration of phosphate in solution is decreased and some phosphate from the active P pool is released. Because the solution P pool is very small, the active P pool is the main source of available P for crops. The ability of the active P pool to replenish the soil solution P pool in a soil is what makes a soil fertile with respect to phosphate. An acre of land may contain several pounds to a few hundred pounds of P in the active P pool. The active P pool will contain inorganic phosphate that is attached (or adsorbed) to small particles in the soil, phosphate that reacted with elements such as calcium or aluminum to form somewhat soluble solids, and organic P that is easily mineralized.

Adsorbed phosphate ions are held on active sites on the surfaces of soil particles. The amount of phosphate adsorbed by soil increases as the amount of phosphate in solution increases and vice versa. Soil particles can act either as a source or a sink of phosphate to the surrounding water depending on conditions. Soil particles with low levels of adsorbed P that are eroded into a body of water with relatively high levels of dissolved phosphate may adsorb phosphate from the water, and vice versa.

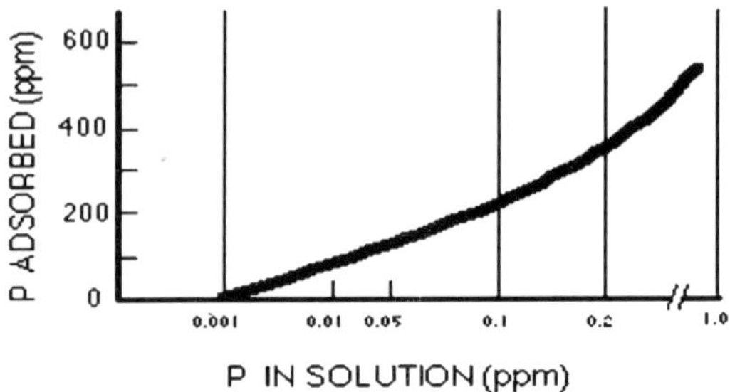


**Fig.** Relationship between P adsorbed by soil and P in solution.

The fixed P pool of phosphate will contain inorganic phosphate compounds that are very insoluble and organic compounds that are resistant to mineralization by microorganisms in the soil. Phosphate in this pool may remain in soils for years without being made available to plants and may have very little impact on the fertility of a soil. The inorganic phosphate compounds in this fixed P pool are more crystalline in their structure and less soluble than those compounds considered to be in the active P pool. Some slow conversion between the fixed P pool and the active P pool does occur in soils.

*Fate of Phosphorus Added to Soils*

The phosphate in fertilizers and manure is initially quite soluble and available. Most phosphate fertilizers have been manufactured by treating rock phosphate (the phosphate-bearing mineral that is mined) with acid to make it more soluble. Manure contains soluble phosphate, organic phosphate, and inorganic phosphate compounds that are quite available. When the fertilizer or manure phosphate comes in contact with the soil, various reactions begin occurring that make the phosphate less soluble and less available. The rates and products of these reactions are dependent on such soil conditions as pH, moisture content, temperature, and the minerals already present in the soil.

As a particle of fertilizer comes in contact with the soil, moisture from the soil will begin dissolving the particle. Dissolving of the fertilizer increases the soluble phosphate in the soil solution around the particle and allows the dissolved phosphate to move a short distance away from the fertilizer particle. Movement is slow but may be increased by rainfall or irrigation water flowing through the soil. As phosphate ions in solution slowly migrate away from the fertilizer particle, most of the phosphate will react with the minerals within the soil. Phosphate ions generally react by adsorbing to soil particles or by combining with elements in the soil such as calcium (Ca), magnesium (Mg), aluminum (Al), and iron (Fe), and forming compounds that are solids. The adsorbed phosphate and the newly formed solids are relatively available to meet crop needs.

Gradually reactions occur in which the adsorbedphosphate and the easily dissolved compounds of phosphate form more insoluble compounds that cause the phosphate to be become fixed and unavailable. Over time this results in a decrease in soil test P. The mechanisms for the changes in phosphate are complex and involve a variety of compounds. In alkaline soils (soil pH greater than 7) Ca is the dominant cation (positive ion) that will react with phosphate. A general sequence of reactions in alkaline soils is the formation of dibasic calcium phosphate dihydrate, octocalcium phosphate, and hydroxyapatite. The formation of each product results in a decrease in solubility and availability of phosphate.

In acidic soils (especially with soil pH less than 5.5) Al is the dominant ion that will react with phosphate. In these soils the first products formed would be amorphous Al and Fe phosphates, as well as some Ca phosphates. The amorphous Al and Fe phosphates gradually change into compounds that resemble crystalline variscite (an Al phosphate) and strengite (an Fe phosphate). Each of these reactions will result in very insoluble compounds of phosphate that are generally not available to plants. Reactions that reduce P availability occur in all ranges of soil pH but can be very pronounced in alkaline soils (pH $> 7.3$) and in acidic soils (pH $< 5.5$). Maintaining soil pH between 6 and 7 will generally result in the most efficient use of phospha.

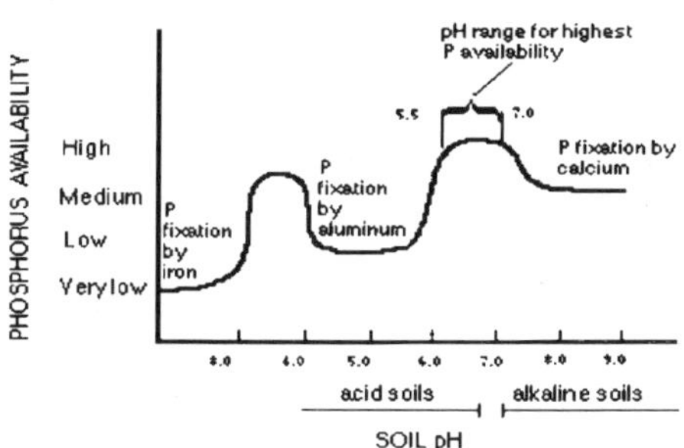


**Fig.** The availability of phosphorus is affected by soil pH.

Adding to the active P pool through fertilization will also increase the amount of fixed P. Depleting the active pool through crop uptake may cause some of the fixed P to slowly become active P. The conversion of available P to fixed P is partially the reason for the low efficiency of P fertilizers. Most of the P fertilizer applied to the soil will not be utilized by the crop in the first season. Continued application of more P than the crops utilize increases the fertility of the soil, but much of the added P becomes fixed and unavailable.

Most fine-to medium-textured soils have large capacities to hold phosphate by adsorption and precipitation. Occasionally the question of how much phosphate a soil can hold is asked, especially when high loading rates of P are expected or have occurred. Soils differ in their phosphate holding capacity. Fine-textured soils can generally hold hundreds of pounds of phosphate per acre.

while coarse-textured soils can generally hold much less phosphate due to the more inert character of sand particles as compared to clay particles. In addition the subsoil of many soils often has an even greater capacity to hold phosphate than does the corresponding surface soil. However, an important aspect of the ability of a soil to hold phosphate is that a soil cannot hold increasing amounts of phosphate in the solid phase without also increasing soil solution phosphate. Increased amounts of phosphate in solution will potentially cause more phosphate to be lost to water running over the soil surface or leaching through the soil. Loading soils with very high levels of phosphate will generally not hurt crops but may result in increased phosphate movement to nearby bodies of water.

*Predicting the Availability of Phosphorus in Soils*

Soils may contain several hundred to several thousand pounds of phosphate per acre. However, much of the phosphate in soils is not available to growing plants. Phosphate in the soil solution P pool is immediately available but the amount is very small in comparison to the total P in soils. The active P pool is phosphorus that can be released into solution but is generally small in comparison to the fixed P.

To determine the need for supplemental P, soil tests are often used to estimate how much phosphate will be available for a crop. The most common way of determining P availability is to mix a small amount of soil with an extracting solution that contains an acid and/or complexing agent that will remove some of the phosphate from the soil particles. The extracting solution and soil are separated by filtration and the amount of P extracted is determined. In alkaline soils, a basic solution may be used as the extractant because an acidic solution will be neutralized by the alkaline soil and be less effective in extracting P. Calibration studies have been done to correlate crop response to fertilizer additions in soils with various soil test levels of P. Using the calibration data, recommendations can be made as to the amounts of phosphate fertilizer that will most likely give optimum yields.

*Soil Phosphorus and Water Quality*

Phosphorus is a somewhat unique pollutant in that it is an essential element, has low solubility, and is not toxic itself, but may have detrimental effects on water quality at quite low concentrations. There is considerable concern about P being lost from soils and transported to nearby streams and lakes. Several chemical properties of soil P have important implications for the potential loss of P to surface water.

Phosphorus in soils is almost entirely associated with soil particles. When soil particles are carried to a river or lake, P will be contained in this sediment. When the sediment reaches a body of water it may act as a sink or a source of

P in solution. In either case, it is a potential source of P that may eventually be released. Most soils have a large capacity to retain P. Even large additions of P will be mostly retained by soils provided there is adequate contact with the soil.

Increasing the amounts of phosphate in soils results in increased levels of phosphate in soil solutions. This will generally result in small but potentially important increases in the amounts of phosphate in water that passes over or through soils.

Phosphate in soils is associated more with fine particles than coarse particles. When soil erosion occurs, more fine particles are removed than coarse particles, causing sediment leaving a soil through erosion to be enriched in P.

*Phosphorus Concerns in the Environment*

Eutrophication can be caused by the nutrient enrichment of a water body. Nutrient movement in run-off and erosion from agricultural non-point sources is a resource management concern. The movement of phosphorus in run-off from agricultural land to surface water can accelerate eutrophication. Undesirable aquatic plant growth results from additions of phosphorus to the water. The net result of the eutrophic condition and excess plant growth in water is the depletion of oxygen in the water due to the heavy oxygen demand by microorganisms as they decompose the organic material. Little attention has been given to management strategies to minimize the non-point movement of P in the landscape because of the easier identification and control of point source inputs of P to surface waters and a lack of direct human health risks associated with eutrophication. Phosphorus is generally the limiting nutrient in fresh water systems and any increase in P is usually results in more aquatic vegetation. Society is concerned about maintaining clean drinking water. This concern now includes a cost for removing the colour, taste, and odour associated with the high trophic condition and vegetation growth in surface water due to excess nutrients.

*Phosphorus Movement in the Landscape*

Phosphorus movement in run-off occurs as particulate P and dissolved P. Particulate P is attached to mineral and organic sediment as it moves with the run-off. Dissolved P is in the water solution. In general, particulate P is the major portion (75-90 per cent) of the P transported in run-off from cultivated land. Dissolved P makes up a larger portion of the total P in run-off from non-cultivated lands such as pastures and fields with reduced tillage. In terms of their impact on eutrophication of water bodies, particulate P becomes less available to algae and plant uptake than dissolved P because of the chemical form it has with the mineral (particularly iron, aluminum, and calcium) and organic compounds. The availability of particulate P to plants and algae is variable, ranging from 10 to 90 per cent of the total P. yet can represent a long-term source of P for algae and plant uptake from the water body. Dissolved P is 100 per cent bioavailable to plants. Added together, the bioavailable portion of particulate P and the dissolved P represents the phosphorus that promotes eutrophication of surface waters.

The method by which P in both particulate and dissolved form moves within the landscape are simplified in the following description. Eroding soil material is transported by run-off. During detachment and movement of sediment in run-off, the finer clay-sized fraction of the source material are preferentially eroded. The P content and reactivity of the eroded material to P are usually greater than the source soil from which it was eroded. The suspended sediment in the run-off can rapidly adsorb the dissolved P in the run-off water.

As run-off moves from the landscape and towards the water body there is generally a progressive dilution of P through additions of water and a reduction in the amount of sediment carried because of sediment deposition. Phosphorus may become more bioavailable by the sorption and Resorption processes and by the preferential transport of claysized material as sediment moves over the landscape.

The movement of dissolved P begins with the Resorption, dissolution, and extraction of P from the soil, plant, and organic material. These processes occur when rain and run-off water interact with the thin layer of surface soil (0.05 to 0.10 inches). Some water infiltrates into the soil and percolates through the profile where desorption of P will result in a low dissolved concentration in subsurface and return flow. High dissolved P concentration can be expected in the water percolating through organic, coarse-textured, and oxygen depleted (reduced), water-logged soils. Soil pH also affects the movement and availability of phosphorus.

The interaction between the particulate and dissolved P in the run-off is very dynamic and the mechanism of transport is complex. Therefore, it is difficult to predict the transformation and ultimate fate of P as it moves through the landscape.

*The Concept*

The purpose of the Phosphorus Index is to provide field staffs, watershed planners, and land users with a tool to assess the various landforms and management practices for potential risk of phosphorus movement to water bodies. The ranking of Phosphorus Index identifies sites where the risk of phosphorus movement may be relatively higher than that of other sites. When the parameters of the index are analyzed, it will become apparent that an individual parameter or parameters may be influencing the index disproportionately. For the first version, the Phosphorus Index will be a 8 x 5 matrix using a limited number of landform site and management characteristics.

The input to the matrix will be readily accessible field data. This index will be used as a tool for understanding the contribution that individual landform and management parameters have towards risk of phosphorus movement and will provide a method for developing management guidelines for phosphorus at the site to lessen their impact on water quality.

The P Index is a simple 8 by 5 matrix that relates site characteristics with a range of value categories. This first version (Table ) of the P Index has eight site characteristics. The eight characteristics are:

- Soil erosion
- Irrigation erosion
- Run-off class
- Soil P test
- P fertilizer application rate
- P fertilizer application method
- Organic P source application rate
- Organic P source application method
- The five value categories are:
- None
- Low
- Medium
- High
- Very high

There are eight site characteristics used in the P Index to assess a particular site. Each site characteristic is rated NONE, LOW, MEDIUM, HIGH, or VERY HIGH by determining the range for each category. Although ranges have been given in this version of the index, it is imperative that each user establish their own range of values for each characteristic. Each user group should consider the cropping, soil, management conditions in the area designated to use the Phosphorus Index. Consideration need to be made as to base conditions for soil erosion, soil phosphorus test levels, and crop requirements. From these local considerations the ranges of the values category can be made. Any user group should include NRCS, Extension Service, University and ARS researchers, as well as state natural resource agencies and other groups interested in water quality.

The definition of each of the eight site characteristics are:

*Soil Erosion*

Soil erosion is defined as the loss of soil along the slope or unsheltered distance caused by the processes of water and wind. Soil erosion is estimated from erosion prediction models currently used (USLE or RUSLE for water erosion and WEQ for wind erosion). Erosion induced by irrigation is calculated by other convenient methods. The value category is given in tons of soil loss per acre per year (ton/acre/year). These soil loss prediction models do not predict sediment transport and delivery to a water body. The prediction models are used in this index to indicate a movement of soil, thus potential for sediment and attached phosphorus movement across the slope or unsheltered distance and towards a water body.

*Irrigation Erosion*

Potential P loss resulting from furrow irrigation induced erosion is considered by inclusion of a rating system based on soil susceptibility to particle detachment by hydraulic shear and flow rate of water in the furrow. The susceptibility to detachment is given by a relative ranking of soil erodibility classes under furrow irrigation (table). These classes are an initial attempt at a relative ranking based on inherent stable and static soil properties (*i.e.*, texture and clay mineralogy).

There are temporal variations in the relative erodibility and actual amount of erosion with furrow erosion. These changes in erodibility are a function in the soil properties or a result of management. However, no attempt is made to consider temporal soil properties nor management factors in the rating. The introduced flow rate in the furrow (Q) is given by the irrigation plan and recorded as gallons per minute (gal/min).

The furrow slope (S) of the site is given as a percentage (feet per 100 feet). The product of flow rate (Q) and slope (S) is used to determine the value category.

*Run-off Class*

The run-off class of the site can be determined from soil survey data. Guidance in determining the run-off class is based on the soil saturated hydraulic conductivity ($K_{Sat}$) and the per cent slope of the site. Another method to determine run-off class is using the NRCS curve number (CN) method. The major factors that determine curve number are the hydrologic soil group. Landscape cover type, conservation treatment, hydrologic condition, and antecedent run-off conditions.

*Soil P Test*

A soil sample from the site is necessary to assess the level of "available P" in the surface layer of the soil. The available P is the level customarily given in a sail test analysis by the Cooperative Extension Service or commercial soil test laboratories. The user of the P Index must determine the ranges of soil test P in each value category (LOW, MEDIUM, HIGH, and EXCESSIVE). These ranges of soil test P values will vary by soil test method and region. The soil test level for "available P" does not ascertain the total P in the surface soil. It does however, give an indication of the amount of total P that may be present because of the general relationship between the forms of P (organic, adsorbed, and labile P) and the solution P available for crop uptake.

*P Fertilizer Application Rate*

The P fertilizer application rate is the amount, in pounds per acre (lb/acre), of phosphate fertilizer ($P_20_5$) that is applied to the soil. This phosphate fertilizer does not include phosphorus from organic sources.

*P Fertilizer Application Method*

The manner in which P fertilizer is applied to the soil and the amount of time that the fertilizer is exposed on the soil surface until crop utilization effects potential P movement. Incorporation implies that the fertilizer P is buried below the soil surface at a minimum of two inches. The value categories of increasing severity, LOW to VERY HIGH, depict the longer surface exposure time between fertilizer application, incorporation, and crop utilization.

*Organic P Source Application Rate*

The organic P application rate is the amount, in pounds per acre (lb/acre), of potential phosphate ($P_2O_5$) that is contained in the manure and applied to the soil. An analysis of the organic material is necessary to determine the potential phosphate content of the manure. The P content by analysis is generally considered to be completely plant available. This organic phosphate source does not include phosphorus from fertilizer sources.

*Organic P Source Application Method*

The manner in which organic P material is applied to the soil and the time that the organic material is exposed on the soil surface until crop utilization can determine potential P movement. Incorporation implies that the organic P material is buried below the soil surface at a minimum of two inches. The value categories of increasing severity, LOW to VERY HIGH, depict the longer surface exposure time between organic P material application, incorporation, and crop utilization.

*The Procedures for Making an Assessment*

The site characteristics have been assigned a weighting based on the reasoning that particular site characteristics may be more prominent than others in allowing potential phosphorus movement from the site. There is scientific basis for concluding that these relative differences exist; however, the absolute weighting factors given are based currently on professional judgement. The site characteristic weighting factors are:

- Soil erosion (1.5)
- Irrigation erosion (1.5)

- Run-off class (0.5)
- Soil P test (1.0)
- P fertilizer application rate (0.75)
- P fertilizer application method (0.5)
- Organic P source application rate (1.0)
- Organic P source application method (1.0)

The value categories are rated using a log base of 2. The greater ratings, the proportionally higher are the values. The higher the value, the higher potential for significant problems related to phosphorus movement.

The value ratings are:

none = 0
low = 1
medium = 2
high = 4
very high = 8

To make an assessment using the P Index, select a rating value for each site characteristic using the categories NONE, LOW, MEDIUM, HIGH, or VERY HIGH. Multiply the site characteristic weight factor by the rating value to get the weighted value for that characteristic. Proceed to rate and factor each characteristic of the index. Sum the weighted values for all eight characteristics, and compare the total with site vulnerability chart.

A description of vulnerability is given to appraise the assessment of the P Index.

### *Sulphur*

Sulphur or sulphur is the chemical element in the periodic table that has the symbol S and atomic number 16. It is an abundant, tasteless, odorless, multivalent non-metal. Sulphur, in its native form, is a yellow crystaline solid. In nature, it can be found as the pure element or as sulfide and sulfate minerals. It is an essential element for life and is found in two amino acids. Its commercial uses are primarily in fertilizers but it is also widely used in gunpowder, matches, insecticides and fungicides.

#### *Notable characteristics*

A piece of sulphur melts to a blood-red liquid. When burned, it emits a blue flame.

At room temperature, sulphur is a soft bright yellow solid. Although sulphur is infamous for its smell—frequently compared to rotten eggs—the odour is actually characteristic of hydrogen sulfide ($H_2S$); elemental sulphur is odorless. It burns with a blue flame that emits sulphur dioxide, notable for its peculiar suffocating odour. Sulphur is insoluble in water but soluble in carbon disulfide and to a lesser extent in other organic solvents such as benzene. Common oxidation states of sulphur include “2, +2, +4 and +6. Sulphur forms stable compounds with all elements except the noble gases.

*Applications*

Sulphuric acid production is the major end use for sulphur, and consumption of sulphuric acid has been regarded as one of the best indices of a nation’s industrial development. Sulphur is also used in batteries, detergents, the vulcanization of rubber, fungicides, and in the manufacture of phosphate fertilizers. Sulfites are used to bleach paper and as a preservative in wine and dried fruit. Because of its flammable nature, sulphur also finds use in matches, gunpowder, and fireworks. Sodium or ammonium thiosulfate are used as photographic fixing agents. Magnesium sulfate, better known as Epsom salts, can be used as a laxative, a bath additive, an exfoliant, or a magnesium supplement for plants. Sulphur is used as the light-generating medium in the rare lighting fixtures known as sulphur lamps.

*Biological role*

The amino acids cysteine and methionine contain sulphur, as do all polypeptides, proteins, and enzymes which contain these amino acids. This makes sulphur a necessary component of all living cells. Disulfide bonds between polypeptides are very important in protein assembly and structure. Homocysteine and taurine are also sulphur containing amino acids but are not coded for by DNA nor are they part of the primary structure of proteins.

Some forms of bacteria use hydrogen sulfide ($H_2S$) in the place of water as the electron donor in a primitive photosynthesis-like process. Sulphur is absorbed by plants via the roots from soil as the sulfate ion and reduced to sulfide before it is incorporated into cysteine and other organic sulphur compounds (sulphur assimilation).

Inorganic sulphur forms a part of iron-sulphur clusters, and sulphur is the bridging ligand in the $Cu_A$ site of cytochrome c oxidase. Sulphur is an important component of coenzyme A.

*Environmental impact*

The burning of coal and petroleum by industry and power plants liberates huge amounts of sulphur dioxide ($SO_2$) which reacts with atmospheric water and oxygen to produce sulphuric acid. This sulphuric acid is a component of acid rain, which lowers the pH of soil and freshwater bodies, resulting in substantial damage to the natural environment and chemical weathering of statues and architecture.

Fuel standards increasingly require sulphur to be extracted from fossil fuels to prevent the formation of acid rain. This extracted sulphur is then refined and represents a large portion of sulphur production.

*History*

Homer mentioned "pest-averting sulphur" in the 8th century BC and in 424 BC, the tribe of Boeotia destroyed the walls of a city by burning a mixture of coal, sulphur, and tar under them. Sometime in the 12th century, the Chinese invented gun powder which is a mixture of potassium nitrate ($KNO_3$), carbon, and sulphur. Early alchemists gave sulphur its own alchemical symbol which was a triangle at the top of a cross. In the late 1770s, Antoine Lavoisier helped convince the scientific community that sulphur was an element and not a compound. In 1867, sulphur was discovered in underground deposits in Louisiana and Texas. The overlying layer of earth was quicksand, prohibiting ordinary mining operations. Therefore the Frasch process was utilized.

*Occurrence*

Elemental sulphur can be found near hot springs and volcanic regions in many parts of the world, especially along the Pacific Ring of Fire. These occurrences are the basis for the traditional name brimstone, since sulphur could be found near the brims of volcanic craters. Such volcanic deposits are currently exploited in Indonesia, Chile, and Japan.

Significant desposits of elemental sulphur also exist in salt domes along the coast of the Gulf of Mexico, and in evaporites in eastern Europe and western Asia. The sulphur in these deposits is believed to come from the action of anaerobic bacteria on sulfate minerals, especially gypsum. Such deposits are the basis for commercial production in the United States, Poland, Russia, Turkmenistan, and Ukraine.

Sulphur extracted from oil, gas and the Athabasca Oil Sands has become a glut on the market, with huge stockpiles of sulphur in existence throughout Alberta.

*Compounds*

Many of the unpleasant odors of organic matter are based on sulphur-containing compounds such as methyl and ethyl mercaptan used to scent natural gas so that leaks are easily detectable. The odour of garlic and "skunk stink" are also caused by sulphur-containing organic compounds. However, not all organic sulphur compounds smell unpleasant; for example, grapefruit mercaptan, a sulphur-containing monoterpenoid is responsible for the characteristic scent of grapefruit.

Polymeric sulphur nitride has metallic properties even though it does not contain any metal atoms. This compound also has unusual electrical and optical properties. This polymer can be made from tetrasulphur tetranitride $S_4N_4$.

*Inorganic sulphur compounds:*

Sulfides ($S^{2-}$), a complex family of compounds usually derived from $S^{2-}$.

Cadmium sulfide (CdS) is an example. Sulfates ($SO_4^{2-}$), the salts of sulphuric acid. Sulphuric acid also reacts with $SO_3$ in equimolar ratios to form pyrosulphuric acid ($H_2S_2O_7$).

Sodium dithionite, $Na_2S_2O_4$, is the highly reducing dianion derived from hyposulphurous/dithionous acid.

Sodium dithionate ($Na_2S_2O_6$).

Polythionic acids ($H_2S_nO_6$), where $n$ can range from 3 to 80.

Sodium polysulfides ($Na_2S_x$)

Sulphur hexafluoride, $SF_6$, a dense gas at ambient conditions, is used as non-reactive and non-toxic propellant

Sulphur nitrides are chain and cyclic compounds containing only S and N. Tetrasulphur tetranitride $S_4N_4$ is an example.

*Isotopes*

Sulphur has 18 isotopes, of which four are stable: $^{32}S$ (95.02 per cent), $^{33}S$ (0.75 per cent), $^{34}S$ (4.21 per cent), and $^{36}S$ (0.02 per cent). Other than $^{35}S$, the radioactive isotopes of sulphur are all short lived. $^{35}S$ is formed from cosmic ray spallation of $^{40}Ar$ in the atmosphere. It has a half-life of 87 days.

When sulfide minerals are precipitated, isotopic equilibration among solids and liquid may cause small differences in the äS-34 values of co-genetic minerals. The differences between minerals can be used to estimate the temperature of equilibration. The äC-13 and äS-34 of coexisting carbonates and sulfides can be used to determine the pH and oxygen fugacity of the ore-bearing fluid during ore formation.

In most forest ecosystems, sulfate is derived mostly from the atmosphere; weathering of ore minerals and evaporites also contribute some sulphur. Sulphur with a distinctive isotopic composition has been used to identify pollution sources, and enriched sulphur has been added as a tracer in hydrologic studies. Differences in the natural abundances can also be used in systems where there is sufficient variation in the $^{34}S$ of ecosystem components. Rocky Mountain lakes thought to be dominated by atmospheric sources of sulfate have been found to have different äS-34 values from lakes believed to be dominated by watershed sources of sulfate.

*Precautions*

Carbon disulfide, carbon oxysulfide, hydrogen sulfide, and sulphur dioxide should all be handled with care.

Although sulphur dioxide is sufficiently safe to be used as a food additive in small amounts, at high concentrations it reacts with moisture to form sulphurous acid which in sufficient quantities may harm the lungs, eyes or other tissues. In creatures without lungs such as insects or plants, it otherwise prevents respiration.

Hydrogen sulfide is quite toxic (more toxic than cyanide). Although very pungent at first, it quickly deadens the sense of smell, so potential victims may be unaware of its presence until it is too late.

*Effect of Sulphur on Wheat Plant Growth*

Sulphur is a building block of protein and a key ingredient in the formation of chlorophyll. Without adequate sulphur, crops can not possibly reach their full potential in terms of yield or protein content. Nor can they make efficient use of nitrogen, phosphorus and other vital elements.

*Sulphur Content and Requirement of Wheat Plants*

Wheat plants have a lesser requirement for sulphur than oilseed crops such as canola. A 2690 kg/ha (40 bu/ac) wheat crop will contain in the seed and straw 12 kg/ha (10 lb/ac) of sulphur, of which only 40 per cent is in the seed.

| | | **Sulphur (S)** | |
|---|---|---|---|
| Yield | Crop Part | Kg/ha | (lb/ac) |
| Wheat | Seed | 5 | 4 |
| 2690 Kg/ha | Straw | 7 | 6 |
| (40 Bu/ac) | Total | 12 | 10 |
| Canola | Seed | 13 | 12 |
| 1960 Kg/ha | Straw | 10 | 9 |

*Time of application*

If sulphate fertilizers are applied in the fall, a portion of the sulphate could leach downward as much as 50 cm (20 in), depending on the amount of fall and spring precipitation and the texture of the soil. Therefore, sulphate fertilizer is usually most efficient when applied in the spring.

Elemental sulphur fertilizer applied in the spring of the crop year may not oxidize to sulphate sufficiently or early enough to supply the crop needs. Farmers who anticipate sulphur deficiencies and plan to utilize elemental S, should apply 10-20 lb/ac of elemental sulphur on an annual to biannual basis to avoid deficiencies.

## Potassium, calcium and magnesium

### *Potassium*

Potassium is a chemical element in the periodic table. It has the symbol K (L. *kalium*) and atomic number 19. The name "potassium" comes from the word "potash", as potassium was first isolated from potash. Potassium is a soft silvery-white metallic alkali metal that occurs naturally bound to other elements in seawater and many minerals. It oxidizes rapidly in air, is very reactive, especially in water, and resembles sodium chemically.

*Calcium*

Calcium is the chemical element in the periodic table that has the symbol Ca and atomic number 20. Calcium is a soft grey alkaline earth metal that is used as a reducing agent in the extraction of thorium, zirconium and uranium. Calcium is also the fifth most abundant element in the Earth's crust. It is essential for living organisms, particularly in cell physiology, and is the most common metal in many animals.

The most abundant isotope, $^{40}Ca$, has a nucleus of 20 protons and 20 neutrons. Its electron configuration is: 2 electrons in the K shell (principal quantum number 1), 8 in the L shell (principal quantum number 2), 8 in the M shell (principal quantum number 3), and 2 in the N shell (principal quantum number 4). The outer shell is the valence shell, with 2 electrons in the lone 4s orbital, the 3p orbitals being empty.

Calcium is not naturally found in its elemental state. Calcium is found mostly in soil systems as limestone, gypsum and fluorite. Stalagmites and stalactites contain calcium carbonate. Being an essential macromineral in the human diet, soil conservation practices often consider the sustainable equilibrium of calcium concentrations in the earth.

All soils contain calcium ions ($Ca_2+$) and magnesium ($Mg_2+$) cations (positively charged ions) attracted to the negative exchange sites on clays and organic matter (cation exchange complex of the soil). The amount and relative proportion usually reflect the soil's parent materials. Calcium (Ca) and Mg are plant-essential nutrients, and the ionic form of each held on the soil exchange sites is the form taken up by plants. The usual approach for determining whether the soil supply is sufficient to meet crop needs is to extract soil with 1 molar (M) ammonium acetate (the same procedure used to determine soil test potassium) and evaluate the amount measured against critical levels. Because Iowa soils contain more than adequate levels of these nutrients, no critical level has been or can be established. Therefore, exchangeable Ca and Mg are not routinely tested, nor are there Iowa State University publications that give soil test Ca or Mg interpretations. Unless you are interested in the soil cation exchange capacity (CEC)—in routine soil testing determined by summing the dominant exchangeable cations ($Ca_2+$, $Mg_2+$, K+, H+)—soil samples don't need to be analyzed in the laboratory for Ca and Mg from most Iowa fields. Also, Iowa soils have large quantities of both nutrients and they are replenished by limestone application.

*How is a Ca:Mg ratio calculated?*

Once exchangeable Ca and Mg are determined by laboratory analysis, the ratio is calculated using the meq basis (electrical charge basis). For example, if there is 4.88 meq Ca/100 g soil and 1.72 meq Mg/100 g, then the Ca:Mg ratio is 2.8:1. Table gives the exchangeable Ca, Mg, and calculated Ca:Mg ratio for several Iowa soils. These values are typical for Iowa soils. Soil Ca:Mg ratios naturally are above 1:1.

*Why the interest in Ca:Mg ratios?*

From the above-mentioned statement that Ca and Mg levels are higher than needed for crop production in Iowa soils, you can easily conclude that ignoring the ratio is just fine. Research confirms that this conclusion is justified; however, promotion of the ratio concept persists today despite many years of research that indicates otherwise. The origin of this concept was derived from work by Bear and colleagues in the 1940s. However, their work did not differentiate between crop response (alfalfa) due to pH improvement from lime application to acid soils and the change in Ca:Mg. Other research during the same time indicated that ratios were not important. Many research trials since have not shown an influence of Ca:Mg ratio on crop production. Conclusions from the researchers were "The results strongly suggest that for maximum crop yields, emphasis should be placed on providing sufficient, but non-excesive levels of each basic cation rather than attempting to attain a favorable basic cation saturation ratio (BCSR), which evidently does not exist." Various greenhouse and field trials indicate that crop productivity is not influenced by ranges from less than 1:1 to more than 25:1—ratios outside of what is normally measured in soils. Plants also play a role in Ca and Mg uptake and exclude excess Ca or Mg at the root surface.

In summary, the Ca:Mg ratio concept is unproven and should not be used as a basis for fertilization or liming practices. Having sufficient levels of Ca and Mg is the proper method of evaluation, rather than trying to manipulate ratios. We are fortunate in Iowa that soil Ca and Mg levels are normally adequate, and maintenance of plant-available Ca and Mg occurs either because the soil has a large supply or because of liming with local quarry limestone to maintain adequate soil pH for crop production.

### *Magnesium*

Magnesium is an essential ingredient of chlorophyll, the green plant pigment that gives leaves their colour and enables the plant to make food from sunlight.

Magnesium is essential for enzymes in plants and especially in chlorophyll – the green pigment that enables plant to carry out photosynthesis. Magnesium is a metallic element represented by the symbol Mg.

*Soil pH*

How well plants grow can be strongly influenced by whether the soil is acid or alkaline. It's best to get the pH right before dealing with other nutrient problems.

Lack of magnesium

This appears as a lack of green in the leaves. The area between the veins of leaves, especially the lower older ones, goes yellow.

Sometimes the leaf margins also lose their green colour. Plants of the cabbage family sometimes develop a purplish colour. Sandy soils are most likely to lack magnesium. This deficiency is also seen when there is plenty of magnesium in the soil, but the roots cannot take it up because of poor soil structure, drought or water-logging. When the underlying problems are resolved the magnesium deficiency isappears.

Fast-growing plants may show a temporary magnesium deficiency in the older leaves. This usually sorts itself out when the plant begins to grow more slowly. There is no need to add extra magnesium. Magnesian limestone is acceptable to organic growers.

*Common magnesium fertilizers*

The amount of magnesium in fertilizer has to be stated on the packet. However, few compound fertilizers contain it, as plants need much less magnesium than nitrogen, phosphorus and potassium. Both Epsom salts (10 per cent Mg) and kieserite (16 per cent Mg) are fast-acting, water-soluble fertilizers containing magnesium sulphate.

Kieserite contains less water in its crystals than the morefrequently sold Epsom salts and is usually cheaper to use when treating large areas. Because it is less soluble, it can be used in seedbeds, whereEpsom salts could scorch delicate young roots.

Epsom salts are used for foliar feeding, as they are more soluble in water than ieserite. Neither fertilizer is organic. Magnesium is easily washed out of soils that contain little or no clay. Since these soils are often acid and need lime, maintain the pH using magnesian limestone. This ground limestone, derived from magnesium-rich rocks, is also known as Maglime or Dolodust. but do not use it where you intend to grow acid-loving plants like rhododendrons and heathers.

Foliar-feed these plants instead –especially rhododendrons which can suffer from magnesium deficiency.

Some liquid fertilizers contain a little magnesium – these are a useful pick-me-up for containergrown plants – tomatoes in growing bags, for example. Manure, seaweed, compost and mushroom compost contain small amounts of magnesium. Cattle manure, for example, contains 0.04 per centMg.

*Magnesium and potassium*

High levels of potassium (K) in the soil suppress plants' uptake of magnesium, so avoid overdosing with fertilizers containing potassium. Conversely, overfeeding with magnesium fertilizer can induce a potassium deficiency in plants. Foliar feeding helps overcome this imbalance.

*General guidelines*

If your soil has not been analysedrecently and you believe that plants are showing magnesium deficiency, try a foliar feed in the first instance. If this has some effect, the soil may be deficient, and applying a magnesium fertilizer should be worthwhile. Spreading 30g a sq m (1oz a sq).

*Suppliers*

Magnesian limestone, Epsom salts and kieserite:

*Bulky organic manures*

Where the magnesium level is low or very low, manure, mushroom compost or garden compost are unlikely to provide enough and you will need fertilizers to boost levels for good growth. Where levels are high or very high, use good-quality manure or compost at 4.5kg a sq m (10lb a sq yd). Manures and composts are highly variable in composition, depending on what they have been made from, how they have been stored and how well-rotted they are. Also, they can vary greatly in water content. If in doubt, consider the organic matter as a soil conditioner and add fertilizer to provide nutrients.

*Foliar feeding*

Wetting the leaves of plants suffering from a magnesium deficiency with a solution of Epsom salts can provide adequate magnesium. However, you may need to foliar feed again during the growing season and every time vulnerable plants are grown. Susceptible perennials, for example, typically need three or four applications per year. Spraying young foliage during warm, humid weather gives the best results. Foliar feeding doesn't work so well on droughtstressed plants with sparsefoliage. Avoid foliar feeding during hot, dry weather as the solution may dry too quickly, and the resulting concentrate could scorch the leaves. If you're in doubt about the suitability of weather conditions, apply the feed in the evening.

Use 20g of Epsom salts per litre of water (3oz per gallon of water). Add a little washing-up liquid – this will act as a 'wetter' to stop the solution running straight off. Check if there is enough wetter by dipping a leaf in the solution – if it runs off without wetting the leaf, more is needed. Too much wetter can damage plants – if you're doing the whole of a precious plant, check by spraying a small shoot and leaving it for a few days before spraying the remainder.

*Soil analysis*

To find out how much magnesium your soil contains requires soil analysis. This should give you a 'magnesium level' from very low to very high, or a 'magnesium index' from 0 to 3. You can then use the tables to determine what to add. Alternatively, a Gardening Which? Soil Analysis will tell you the levels and recommend what action to take.

### *Trace Element*

Trace element is a chemical element whose concentration is less than 1000 ppm or 0.1 per cent of rocks composition.

*Microminerals*

Microminerals(also known as trace elements) are micronutrients that are chemical elements. They include boron, chromium, cobalt, copper, fluorine, iodine, iron, manganese, molybdenum, selenium, silicon, tin, vanadium, and zinc. They are dietary minerals needed by the human body in very small quantities (generally less than 100mg per day) as opposed to macrominerals which are required in larger quantities. Note that the use of the term "mineral" in this context is to be distinguished from the definition commonly used in the geological sciences.

*Micronutrients for plants:*

There are about eight nutrients essential to plant growth and health that are only present in very small quantities. These are manganese, boron, copper, iron, chlorine, cobalt, molybdenum, and zinc. Some consider sulphur a micronutrient, but it is listed here as a macronutrient. Though these are present in only small quantities, they are all necessary.

Boron - One of eight micronutrients. Boron is believed to be involved in carbohydrate transport in plants; it also assists in metabolic regulation. Boron deficiency will often result in bud dieback.

Chlorine - One of eight micronutrients in plants. Chlorine is necessary for osmosis and ionic balance; it also plays a role in photosynthesis.

Cobalt - One of eight micronutrients essential to plant health. Cobalt is thought to be an important catalyst in nitrogen fixation. It may need to be added to some soils before seeding legumes.

Copper - One of eight micronutrients in plants. It is a component of some enzymes and of vitamin A. Symptoms of copper deficiency include browning of leaf tips and chlorosis.

Iron - One of eight micronutrients in plants. It is essential for chlorophyll synthesis, which is why an iron deficiency results in chlorosis.

Manganese - One of eight micronutrients. Manganese activates some important enzymes involved in chlorophyll formation. Manganese deficient plants will develop chlorosis between the veins of its leaves. The availability of manganese is partially dependent on soil pH.

Molybdenum - One of eight micronutrients essential to plant health. Molybdenum is used by plants to reduce nitrates into usable forms. Some plants use it for nitrogen fixation, thus it may need to be added to some soils before seeding legumes.

Zinc - One of eight micronutrients necessary for plant health. Zinc participates in chlorophyll formation, and also activates many enzymes. Symptoms of zinc deficiency include chlorosis and stunted growth.

Micronutrients are essential elements only needed by life in small quantities. Vitamins and trace minerals are sometimes included in the term.

Microminerals or trace elements include at least iron, cobalt, chromium, copper, iodine, manganese, selenium, zinc, and molybdenum. They are dietary minerals needed by the human body in very small quantities (generally less than 100mg/day) as opposed to macrominerals which are required in larger quantities.

Vitamins are organic chemicals that a given living organism requires in trace quantities for good health, but which the organism cannot synthesize, and therefore must obtain from its diet.

*Macronutrients*

Macronutrients can be broken into two more groups: primary and secondary nutrients.

The primary nutrients are nitrogen (N), phosphorus (P), and potassium (K). These major nutrients usually are lacking from the soil first because plants use large amounts for their growth and survival.

The secondary nutrients are calcium (Ca), magnesium (Mg), and sulphur (S). There are usually enough of these nutrients in the soil so fertilization is not always needed. Also, large amounts of Calcium and Magnesium are added when lime is applied to acidic soils. Sulphur is usually found in sufficient amounts from the slow decomposition of soil organic matter, an important reason for not throwing out grass clippings and leaves.

*Nitrogen (N)*

Nitrogen is a part of all living cells and is a necessary part of all proteins, enzymes and metabolic processes involved in the synthesis and transfer of energy.

Nitrogen is a part of chlorophyll, the green pigment of the plant that is responsible for photosynthesis.

Helps plants with rapid growth, increasing seed and fruit production and improving the quality of leaf and forage crops.

Nitrogen often comes from fertilizer application and from the air (legumes get their N from the atmosphere, water or rainfall contributes very little nitrogen)

*Phosphorus (P)*

Like nitrogen, phosphorus (P) is an essential part of the process of photosynthesis.

Involved in the formation of all oils, sugars, starches, etc.

Helps with the transformation of solar energy into chemical energy; proper plant maturation; withstanding stress.

Effects rapid growth.

Encourages blooming and root growth.

Phosphorus often comes from fertilizer, bone meal, and superphosphate.

*Potassium (K)*

Potassium is absorbed by plants in larger amounts than any other mineral element except nitrogen and, in some cases, calcium.

Helps in the building of protein, photosynthesis, fruit quality and reduction of diseases.

Potassium is supplied to plants by soil minerals, organic materials, and fertilizer.

*Calcium (Ca)*

Calcium, an essential part of plant cell wall structure, provides for normal transport and retention of other elements as well as strength in the plant. It is also thought to counteract the effect of alkali salts and organic acids within a plant.

*Magnesium (Mg)*

Magnesium is part of the chlorophyll in all green plants and essential for photosynthesis. It also helps activate many plant enzymes needed for growth.

Soil minerals, organic material, fertilizers, and dolomitic limestone are sources of magnesium for plants.

*Sulphur (S)*

Essential plant food for production of protein.

Promotes activity and development of enzymes and vitamins.

Helps in chlorophyll formation.

Improves root growth and seed production.

Helps with vigorous plant growth and resistance to cold.

Sulphur may be supplied to the soil from rainwater. It is also added in some fertilizers as an impurity, especially the lower grade fertilizers. The use of gypsum also increases soil sulphur levels.

## FOREST ECOSYSTEMS

### Term of Ecosystem

A system is a group of parts that interact through one or more processes. The term ecosystem was introduced and defined by Tansley (1935), who as "a fundamental organizational unit of the natural world that includes both organisms and their spatial environment." Ecosystems have since been defined in various ways, and at different spatial and temporal scales. Some ecologists define ecosystems on the basis of biotic organisms, populations, or communities. For example, Hutchinson (1978) considered the ecosystem to be the environmental context in which population or community dynamics occur. Others define ecosystems in terms of their abiotic characteristics and processes. For example, Lindeman (1942) defined ecosystems as "...the system composed of physical, chemical, and biological processes active within a space/time unit." Regardless of whether the emphasis is on biotic components or abiotic characteristics and processes of ecosystems, both remain integral to the concept of ecosystem. Rowe (1961) emphasized this when he defined ecosystems as "...a three dimensional segment of the earth where life forms and the environment interact."

Wetland ecosystems have been defined in a variety of ways by researchers, resource managers, and regulatory authorities, depending on their specific needs and objectives. In the applied world of regulation, planning, and management, wetlands are usually defined in terms of their physical, chemical, and biological characteristics such as hydrologic regime, soil type, and plant species composition. For example, in classifying wetlands for mapping, inventory, and other purposes, Cowardin *et al.* (1979) defined wetlands as "...lands transitional between terrestrial and aquatic systems where the water table is usually at or near the surface or the land is covered by shallow water..." that are characterized by the presence of hydrophytic vegetation, hydric soils, and surface water during the growing season. Wetlands are often biodiversity 'hotspots', as well as functioning as filters for pollutants from both point and non-point sources, and being important for carbon sequestration and emissions.

The value of the world's wetlands are increasingly receiving due attention as they contribute to a healthy environment in many ways. Wetland functions are defined as the normal or characteristic activities that take place in wetland ecosystems or simply the things that wetlands do. Wetlands perform a wide variety of functions in a hierarchy from simple to complex as a result of their physical, chemical, and biological attributes. For example, the reduction of nitrate to gaseous nitrogen is a relatively simple function performed by wetlands when aerobic and anaerobic conditions exist in the presence of denitrifying bacteria.

Nitrogen cycling and nutrient cycling represent increasingly more complex wetland functions that involve a greater number of structural components and processes. At the highest level of this hierarchy is the maintenance of ecological integrity, the function that encompasses all of the structural components and processes in a wetland ecosystem. Wetlands are one of the most productive of all ecosystems, and carry out critical regulatory functions of hydrological processes within watersheds. Regulating water quality, water levels, flooding regimes, and nutrient and sedimentation levels are a few of these processes. As with any natural habitat, wetlands are important in supporting species diversity and have a complex of wetland values. Moreover, the pattern of seasonal variation of the wetland affects the bird population fluctuation. Even small wetlands are extremely important to the conservation of biodiversity because they provide critical breeding habitat where dispersed populations can exchange genetic material, reducing the risks of extinction.

The present review is aimed at providing in a nutshell, the distribution of wetlands, the value of Wetlands, the causes and consequences of the loss of wetlands and their conservation status with special reference to India.

## Management of Ecosystem

Practical ecosystem management usually involves maintaining a natural or artificial homeostasis in which the end product is of benefit to man, and in which violent or pathologic imbalances can be controlled. An example of a natural ecosystem which man might wish to maintain for esthetic or recreational value would be a wilderness area or National Park. Here the most important consideration is to insure that natural processes are permitted to continue. In a forest this would mean, for example, that fire, if already a natural part of the ecosystem, be permitted to occur at natural frequency and intensity.

Several recent studies have shown that certain forests depend upon intermittent burning to recycle nutrients and minerals. In the lowland conifer forests of Alaska, if fire is excluded a thick carpet of moss may accumulate and raise the permafrost level. This encourages the growth of black spruce, a species with little food or timber value. Browsing animals such as deer, moose and bear cannot survive on black spruce. They depend on the Tarai fires to maintain a suitable cycle of vegetation. Many foresters now feel that the underbrush and leaf litter in the Ponderosa pine forests of western United States should be removed by light ground fires every few years to maintain soil fertility, to prevent excessive accumulation of organic matter in the form of cellulose and also to prevent more devastating fires that destroy the entire forest.

In the pine forests of the South, very efficient fire suppression techniques allowed dangerous accumulation of combustible materials, and a pine fungus disease was favoured in the accumulated litter. In all of these forests, proper management should not exclude all fires, but should prevent totally devastating fires. These examples illustrate the need for understanding natural processes in ecosystems in order to maintain them properly.

The Everglades of Florida provide another example of the importance of understanding natural ecosystem processes before the system can be properly managed or maintained. The unique flora and fauna of the Everglades depends upon the proper balance of fresh and saline water flowing through its channels. If a significant decline in fresh water input occurs, as has been taking place due to the diversion of fresh water to the growing megalopolis of Miami, numerous important species of plants and animals may die and the characteristic features of the system may be destroyed. In times of drought, many animals depend upon alligator wallows as a source of water. These are pools hollowed out by alligators, and they tend to retain water longer than other parts of the swamp. If alligators decline through shooting or any other.

It is increasingly important to understand the dynamics of ecosystem, their patterns of succession and evolution, and their patterns of succession and evolution, and their patterns of diversity and stability. E.P. Odum has written a major review of these topics and has shown that ecosystem may be characterized as pioneer, developmental and mature. Various stages within this continuum possess certain propertics of production, respiration, community structure, nutrient cycling, homeostasis, etc. For example, young developmental stages of an ecosystem tend to have greater gross production in relation to respiration, but lower species diversity and less stability than mature stages.

Thus, developmental stage to be favoured in agricultural practice, but they are also more vulnerable to ecologic damage or catastrophe. We will explore these relationships in more detail throughout subsequent chapters after various components of the ecosystem are considered seperately, diversity and stability. E.P. Odum has written a major review of these topics and has shown that ecosystem may be characterized as pioneer, developmental and mature.

Various stages within this continuum possess certain properties of production, respiration, community structure, nutrient cycling, homeostasis, etc. For example, young developmental stages of an ecosystem tend to have greater gross production in relation to respiration, but lower species diversity and less stability than mature stages. Thus, developmental stage to be favoured in agricultural practice, but they are also more vulnerable to ecologic damage or catastrophe. We will explore these relationships in more detail throughout subsequent chapters after various components of the ecosystem are considered seperately.

**The Principles of Ecosystem Management**

Even if some synergies between the different international arrangements and processes relating to the biodiversity of freshwater resources still deserve more attention and none of the international treaties and institutions already deals fully with freshwater ecosystems, there is an increasing consensus on the highest political levels to manage water, water-related processes, and ecosystems in a sustainable manner. However, at the level of implementation the value of ecosystems is rarely taken fully into account. In order to close this implementation gap the contracting parties of the CBD and other advocates have proposed a so-called "ecosystem-based approach" as an appropriate way of managing natural resources.

## The Ecosystem Approach

The ecosystem approach is a strategy for the integrated management of land, water and living resources that promotes conservation and sustainable use in an equitable way. The term ecosystem is used more as a mental construct suggesting complexity and systems interaction rather than a geographic entity. An emphasis on the complexity of system-wide interactions highlights scientific uncertainty and results in the need to deal with this uncertainty explicitly by acting conservatively and managing adaptively: setting a course of action based on a set of hypotheses, monitoring what happens, and re-evaluating the direction on what one learns. At its most fundamental level, an ecosystem approach maintains diversity as a means of building resilience against catastrophic events in biological, economic, organisational, and political systems. At its second meeting in 1995, the Conference of the Parties of the CBD adopted the ecosystem approach as the primary framework for action, and subsequently has referred to the ecosystem approach in the elaboration and implementation of the various thematic and cross-cutting issues work programmes under the Convention. The issues concerned include:

- Biological diversity of inland water ecosystems.
- Marine and coastal biological diversity.
- Agricultural biological diversity.
- Forest biological diversity.
- Indicators of biological diversity.
- Incentive measures and environmental impact assessment.

By recommending the ecosystem approach as the guiding concept, in May 2000 the fifth COP took a next step towards its practical verification in the frame of the CBD.

Concerning inland water ecosystems a work programme was adopted by the COP which highlights the importance of designing integrated watershed, catchment and river basin management strategies. Particularly the interrelation between water management and agriculture was stressed, as well as the need to use the ecosystem approach in all sectors which have an impact on inland water biodiversity.

The ecosystem approach was initially developed as a research paradigm and not intended to serve as the basis for resource management. With the political application of the notion "ecosystem-based approach" advocates intended to find an adequate language in response to the shortcomings of classical nature conservation approaches.

Although classical nature conservation (*e.g.* designation of protected areas) can be considered successful in many regions and also is an important module in ecosystem management, its recognition of human needs and other legitimate sectoral interests was rather insufficient. It did not put enough emphasis on the biological diversity outside protected areas. Furthermore, the primary method of protecting freshwater biodiversity has been to designate particular species as threatened or endangered making them subject to national recovery programmes or international protection. Unfortunately, this approach is failing and it demonstrates the need for a more holistic approach in order to overcome the shortcomings of traditional efforts in nature protection policy. In the United States, for example, no aquatic species has ever graduated from the government's endangered species list, but 10 species of fish have been removed due to extinction. Equally, ecosystem management responds to the obvious shortcomings of classical single-resource approaches focused on exploitation.

**The 12 Malawi Principles**

- Management objectives are a matter of societal choice.
- Management should be decentralised to the lowest appropriate level.
- Ecosystem managers should consider the effects (actual or potential) of their activities on adjacent and other ecosystems.
- Recognising potential gains from management there is a need to understand the ecosystem in an economic context. Any ecosystem management programme should:
  - Reduce those market distortions that adversely affect biological diversity.
  - Align incentives to promote sustainable use.
  - Internalise costs and benefits in the given ecosystem to the extent feasible.
- A key feature of the ecosystem approach includes conservation of ecosystem structure and functioning.
- Ecosystems must be managed within the limits to their functioning.
- The ecosystem approach should be undertaken at the appropriate scale.
- Recognising the varying temporal scales and lag effects which characterise ecosystem processes, objectives for ecosystem management should be set for the long term.
- Management must recognise that change is inevitable.
- The ecosystem approach should seek the appropriate balance between conservation and use of biological diversity.
- The ecosystem approach should consider all forms of relevant information, including scientific and indigenous and local knowledge, innovations and practices.
- The ecosystem approach should involve all relevant sectors of society and scientific disciplines. These principles were agreed at CBD's COP-5 under decision V/6 with minor modifications.

Therefore, ecosystem management is proposed as a modern way of managing natural systems and generally intends to use natural resources within their natural limits. However, although there are still some controversies on how to translate ecosystem management into concrete management terms, parties of the CBD have already agreed on basic principles to guide the *modus operandi*.

A key point is that the ecosystem approach is a holistic process for integrating and delivering in a balanced way the three objectives of the CBD: conservation and sustainable use of biological diversity combined with equitable sharing of biodiversity's benefits. In particular, the agreed principles of the ecosystem approach stress the importance of:

- Societal choice concerning management objectives.
- The "economics" of biodiversity (internalisation of costs and benefits, balance between conservation and use).
- Adequate institutional structures (decentralisation, management at the appropriate scale).
- Adaptive management which enables policy makers to anticipate and cater for the natural dynamics of ecosystems.
- Consideration not only of scientific knowledge but also of indigenous and local knowledge.

In order to translate this rather abstract principles into guidelines for water management it is first necessary to define the appropriate management level to meet the general objectives. While the cited principles stress the importance of decentralised approaches in general terms, the idea of river basin management is considered an adequate starting point for the implementation of the ecosystem approach in water management.

The reasoning is that river basins offer many advantages for strategic planning and that they represent the most appropriate geomorphologic units on which to base ecosystem management practices. However, concerning ecosystem protection there are some limits: groundwater catchments only rarely match with surface water catchments; water quality and quantity inside the river basin interacts with other influences outside the catchments boundaries; estuarine ecosystems, for example, can only be sufficiently protected by taking into account the interaction of freshwater and tidal currents.

Therefore, river basin management should not be perceived as a panacea but it does provide a sound geographical basis for implementation. Briefly, ecosystem management should be an integral component of comprehensive river basin management strategies. Hence, the explicit switch to ecosystem management leads to an accentuation of the main focus in water management because ecosystem management demands:

- A broader perspective: whole ecosystems and the full range of services, no sectoral view.
- Efforts to balance natural ecosystem robustness and human-induced alterations.

- Collation of information on the status of, and threats to, the biodiversity of the freshwater ecosystem concerned (including impacts of the introduction of alien species).
- Explicit consideration of land-use, in-stream flow and water temperature for maintaining ecological functions.
- A long-term perspective.
- Consideration of the full economic value of freshwater ecosystems (include its biodiversity).
- Recognition of ecosystem's change as a matter of fact (*i.e.* restoration of damaged or destroyed ecosystems will never lead to the "pristine" state; adaptive management to deal with uncertainties).
- Integration of scientific knowledge and modern management tools (Decision Support Systems, Geographical Information Systems etc.) into decision-making.
- Identification of win-win options between protection and use of freshwater ecosystems: "use them, but don't lose them".
- Consideration of the full range of policy instruments applicable in water management (economic and legal instrument, co-operation etc.)

As most of these elements are widely agreed upon the rhetoric of international water policy, the next steps must include attempts to tackle the implications of the ecosystem approach in practice. For example, only little is known about the value of ecosystem management in political terms for at least two reasons:

- The concept appears highly demanding because much information – sometimes on "invisible" relationships between human-induced pressures and natural reactions – is needed, and effective vertical and horizontal co-ordination of sector policies and planning instruments has to take place. Today, even in wealthy nations a lack of integration across sectors in managing water resources can be identified.
- It is necessary to decide which services of water receive priority, and how the costs and benefits are to be distributed. The developed world, for example, can "afford" to stop or to remove development projects in favour of ecosystem integrity (*e.g.* Great Whale Hydropower Project in Canada, Edwards Dam in the US) and to use sometimes generous subsidies to compensate land owners for restricting their use of the lands. The restoration of river ecosystems in the developed world also is an example for costs emerging in the short terms. Although long-term economic benefits are regularly provided, political decision are often dominated by short-term considerations.

**Water and Forested Ecosystems**

Ecologists consider water to be the defining part in an ecosystem, including the forest ecosystem. Water shapes the physical landscape through erosion Undulating forest surrounds meadows, lakes, and rivers of the Charlevoix Conservation Area, a coastal watershed in Quebec, Canada. This region, which illustrates a high-quality forested watershed, has been designated a biosphere by the United Nations Educational, Scientific and Cultural Organization (UNESCO). and deposition. It also shapes the biological parts of the ecosystem by its presence or absence; its quantity and quality; and its occurrence and distribution. The water cycle plays a key role in ecosystem functions and processes. Forests, in turn, are vital to the water cycle and to water quality. In essence, the forest acts like a giant sponge, filtering and recycling water. Approximately 80 per cent of U.S. fresh-water resources are estimated to originate in forests, which cover one-third of the U.S. land area.

Tree leaves intercept water from rain, snow, and fog; the leaves also release water back to the atmosphere by evapotranspiration. Tree roots extract water from the soil while helping hold the soil in place. Forested land reduces the surface impact of falling rain through interception and delay of water reaching the surface.

Forestland also decreases the amount and velocity of storm run-off over the land surface. This in turn increases the amount of water that soaks into the ground, a portion of which can ultimately recharge underlying aquifers. Conversely, water from hydraulically connected surficial aquifers may enter streams and wetlands, helping to maintain their water levels during dry periods.

**Forests and the Hydrologic Cycle**

The surface water in a stream, lake, or wetland is most commonly precipitation that has run off the land or flowed through topsoils to subsequently enter the waterbody. If a surficial aquifer is present and hydraulically connected to a surface-water body, the aquifer can sustain surface flow by releasing water to it. In general, a heavy rainfall causes a temporary and relatively rapid increase in streamflow due to surface run-off. This increased flow is followed by a relatively slow decline back to baseflow, which is the amount of streamflow derived largely or entirely from groundwater.

During long dry spells, streams with a baseflow component will keep flowing, whereas streams relying totally on precipitation will cease flowing. Generally speaking, a natural, expansive forest environment can enhance and sustain relationships in the water cycle because there are less human modifications to interfere with its components. A forested watershed helps moderate storm flows by increasing infiltration and reducing overland run-off. Further, a forest helps sustain streamflow by reducing evaporation (*e.g.*, owing to slightly lower temperatures in shaded areas). Forests can help increase recharge to aquifers by allowing more precipitation to infiltrate the soil, as opposed to rapidly running off the land to a downslope area.

### Riparian Areas

The riparian zone is broadly defined as the area between a body of water and the upland parts of the landscape that are rarely flooded except under the most extreme conditions. But the term also can refer more specifically to the immediate streamside area. Riparian areas represent less than 10 per cent of most forest ecosystems, yet these areas often are the most productive portions. Compared to upland regions, riparian areas have more water available; the vegetation is more robust; the soils are deeper; the timber often is of higher quality; and the waterbodies have more shade. The riparian zone also may include wetlands bordering streams and lakes. This combination of factors makes riparian areas among the most heavily used portions of a forest. Riparian and wetland areas provide abundant and reliable forage for wildlife, as well as transportation corridors. They also may receive heavy human use for recreation.

Riparian zones also are attractive destinations for logging and for livestock grazing; as a result, riparian areas in forests are sometimes heavily damaged, especially in the forests of the arid American Southwest. Fortunately, riparian areas respond well to good management practices.

### Aquatic Biodiversity

Forest lands and waters are vitally important in maintaining biodiversity and providing habitat for fish and wildlife, including threatened or endangered aquatic species. In the United States, over one-third of national forest lands are critical for maintaining aquatic biodiversity and protection of listed species. For aquatic species, watersheds provide the basic unit of any conservation strategy. Many watersheds also contain isolated habitats with unique characteristics producing a high potential for rare species. Some species occur only near a single spring or in a single stream within a given watershed. Lands set aside to protect these unique habitats also benefit the entire watershed and its ecosystem.

### Development of Community Wastelands and Denuded Forests

Conservation of biodiverstity and protection of forests are the other important components of community forestry. However there has been a lukewarm response to these activities in the past, mainly because of lack of awareness, inadequate benefits and lack of trust between the local communities and government agencies who own these lands. Hence the new strategy should find a suitable solution for these problems.

*Development of forest lands*: Rehabilitation of denuded forest lands and protection of natural forests are essential to conserve biodiversity. The forests influence the climate and soil moisture conditions and improve agricultural production.

This aspect needs wider publicity in rural areas and the local people should be motivated to take active part in Joint Forest Management. The Village Forest Committees willing to protect the natural forests, should be encouraged to share the benefits in the form of fodder, fuel, timber and NWFP. These committees can also take up reforestation of degraded forest lands, aiming to conserve biodiversity and generate income. While facilitating natural regeneration, the local communities can also introduce fuel, timber and NWFP species of their choice to enhance the productivity on denuded forest lands.

*Development of community lands*: Lack of people's initiative was a serious problem in developing community wastelands, mainly due to lack of direct benefits. Earlier efforts to establish woodlots could not generate substantial income. Establishment of NWFP species such as tamarind (*Tamarindus indica*), Indian gooseberry (*Emblica officinalis*), neem (*Azadirachta indica*), Pongamia (*Derris indica*), *Madhuca spp*. and a wide range of medicinal species can generate substantial income, regularly for over 50-100 years. With the world's herbal medicine business likely to cross US $ 3 trillion during the next decade, it will be an excellent opportunity for the Asian farmers to enhance their income through cultivation of these species on community lands as well as on private wastelands.

*Development of Mined Areas*: Rehabilitation of mined areas is another cause of serious concern. Although the industries are obliged to establish a green cover after mining the area, the outcome has been very poor. Such sites are major causes of soil erosion and flooding of rivers. Lack of commitment, inadequate expertise in selection of plant species and their management have been the main reasons for failure. These activities can be entrusted to voluntary organisations and local communities, with research support to raise healthy plantations. Preference should be given to NWFP and timber species which have long gestation period, sufficient to enrich the mined lands.

*$CO_2$ Sinks*: There are opportunities for environmental protection by establishing industrial greenbelts, recycling treated effluent and developing carbon dioxide sinks, particularly around urban areas. The polluting industries can be compelled to establish greenery through the involvement of voluntary organisations and local people.

*Environmental Awareness and Eco-tourism*: Lack of awareness among the common public is a cause for environmental pollution in the developing countries. Hence, continuous efforts are needed to motivate the common public to plant more trees and protect the environment. School children and literate urban population are fairly receptive to participate in environmental protection. This can be undertaken through school based plant nurseries, development of arboretum, botanical gardens and greenbelts around the schools and industries.

While developing industrial greenbelts and afforestation on river banks, special efforts can be made to promote eco-tourism. Creating waterbodies through plugging of gullies, developing, nature trails, shady grounds for resting, selecting suitable plant species to attract wildlife and providing adequate eco-tourism. Environmental exhibitions at these sites can help in motivating the public to take active part in community forestry and environmental protection.

**Organisational Development**

Development of local organisations is the key to ensure people's participation and to sustain the benefits of investments and efforts made in community forestry.

**Tree Growers Organisations**

The initial need is to provide leadership to organise local people in afforestation. This can be carried out through formation of local Selp-Help Groups, village level farmers' associations and forest groups. This activity should receive priority during the next decade.

**Capacity Building**

Training of local farmers, supply of critical inputs, credit and technical advisory services to improve the profitability and developing market networks are the critical inputs which have to be introduced in the field to support tree growers.

**Networking of Tree Growers and User Groups**

There is an urgent need to set up networks of tree growers, forestry users and concerned scientists in these fields. Such networks should identify the opportunities for tree growers. They can also promote processing and marketing of forestry products and protect the interests of tree growers.

**Forestry Information Service**

Setting up of information centres to identify the priority of the industries and farmers and provide necessary information to enhance the economic viability of community forestry would help to sustain the interest of tree growers. Forestry Extension Units should closely associate with Information Service Centres to identify appropriate technologies for transferring to the field.

**Forestry Research**

The networks should identify the problems affecting the production and marketing of the produce and encourage the Forestry Research Institutions to take up relevant scientific studies to solve the problems. The researchers should closely associate with the network and Information Centres to disseminate new technologies to solve field problems and enhance the production.

Forestry research should focus on the following topics:

- Conservation of biodiversity;
- Improving the productivity of wastelands, tree plantations and natural forests;
- Exploring the utility of multipurpose tree species and NWFP;
- Developing of alternatives for conserving wood;
- Selecting various organisational systems to manage community forestry.

The outcome of these research topics can further strengthen the community forestry programme in the country.

**Invertebrate Species in Forest Ecosystems**

The diversity and abundance of invertebrate species in forest ecosystems are determined by the interaction of complex ecological and environmental factors. Research into these factors is gradually providing data that can be incorporated into models, both descriptive and mathematical, of the processes involved.

At the core of such models is the description of the availability of resources that enables a particular invertebrate species to reproduce, whereas the rate of reproduction is dependent on multiple factors, including temperature, competition, etc., that determine utilization of available resources. In relation to exploitation of resources within a forest ecosystem, the primary focus for forest managers is the potential of invertebrates to cause damage and to reduce yields of the trees themselves. This is certainly the case in managed forests where the primary aim is to grow trees for commercial gain, often with associated value from recreational, landscape or environmental considerations.

Any invertebrates that feed directly on the trees, or have an influence of how the trees grow, may be classified as pests and hence may require some form of pest management to reduce the threat. But it is pertinent to consider how ecological theory and practice can be used both to understand why a pest problem has arisen and to provide guidance on how to combat that problem. The same principles can be applied to understanding and enhancing the biodiversity of forests, including conservation of rare and beneficial invertebrates. Indeed, the two extremes of pest outbreak and invertebrate rarity are part of the same continuum of interaction between invertebrates and their host trees.

In general, considering that trees are known to harbour the greatest numbers of insect species (diversity), it is surprising that serious pest status is reached only relatively infrequently. For example, although oak *(Quercus* spp.) is known to support at least 450 species of insect in Britain, Bevan (1987) lists only 35 species as pests, of which five were classified as important or severe (XXX-XXXXX on Bevan's scale). Wider listing of pest species has been noted for forest insects in continental Europe, although even here the number of consistently serious pests is small relative to the total diversity of insects present (Klimetzek 1993).

Naturally, the severity of a pest outbreak will be classified differently depending on the purposes of the forest manager, and thus it is not always possible to classify insects consistently as pests across geographical ranges or, indeed, between years. Further, an insect may normally be classified as uncommon or rare but, given an abundance of resources, may be described later as a pest. This was certainly the case for oak pinhole borer, *Platypus cylindrus* (Coleoptera: Platypodidae), which prior to the 1987 great storm in southern England was listed in the British Red Data Book as rare.

Winter (1988) warned that the great increase in availability of freshly dead oak following the storm could result in an increase in numbers of *P. cylindrus* and, subsequently, it has been noted that the insect is now causing damage to high-value oak logs as a result of its habit of boring directly into the heartwood. Therefore, consideration of why invertebrate abundance might increase above a given threshold level can provide a more consistent appraisal of status and also potential approaches for reducing populations below that threshold or indicating measures that can be taken to encourage rare or endangered species.

Among the many interacting processes that determine insect diversity and abundance, a few attributes that are particularly well linked to the environmental status of trees can be distinguished. These provide a means to describe intrinsic processes but also allow comparison with other plants, thus offering the potential for a more consistent approach to descriptive population ecology of invertebrates irrespective of the host plant being considered.

**Longevity**

Trees are the longest-lived components of an ecosystem and, in the majority of natural situations, approach or represent the climax stage of vegetation succession. This provides opportunities for continuity of insect population growth and over time this will give a particular set of insect associates for a given tree species.

Continuing the theme of resource availability, the fact that trees may be present in the same location for tens or hundreds of years ensures that both migrant and, particularly, relatively sedentary invertebrates can exploit them at some stage during their growth. However, longevity *per se* is not a quantifiable attribute but does provide the means for the other components, to influence invertebrate diversity and abundance. Increasing stability in the environment, as represented under forest conditions, leads to stability in both microclimate and resource availability, particularly favouring exploitation by specialist insect feeders and their associates.

**Biogeography**

Tree species richness in a given region or locality varies considerably with geographical location and thus the climax community structure and longevity of individual species may vary accordingly. Although subject to uncertainty and lack of unequivocal evidence of the underlying processes, there is certainly a gradient of species richness of plant cover, especially trees, from the tropics to the poles. Not surprisingly, the same applies to a wide range of animals and plants; as an extreme example, the species richness of orchids ranges from 15 to 2500 in a latitudinal range from 0 to 66° S.

Further examples are provided by Brown (1988). More specifically, in an analysis of species richness of deciduous forests both in the tropics and, latitudinally, within North America, Brown and Gibson (1983) contrasted the high diversity of tree species in the tropics (up to 100 $ha^{-1}$) with the low diversity from north to south within North America (1–30$ha^{-1}$). However, even within such a generalization there are exceptions, so that for example conifers are more diverse in temperate latitudes.

**Geological Time**

The ecological structure of forests and of their associated invertebrate faunas is determined by both present and recent characteristics, particularly area, and by their history in geological time. The latter characteristic is driven as much by *time per se* as by the history of major disturbances (glaciations, volcanic eruptions, land shifts, etc.) within geological time. Thus, present flora and fauna may well be determined by recolonization since the last major disturbance, the diversity of species being a reflection of both colonization and speciation relative to more recent ecological conditions.

The gradient of species from polar to tropical regions could therefore be explained by the high level of disturbance closer to the polar masses compared with the relative stability of climate in the tropics. Evidence from habitats that have become isolated since the last major disturbance indicates that current species diversity may reflect the historical link to the original major habitat. For example, islands that were once joined by land bridges to the larger land masses of New Guinea have species diversities of birds greater than expected compared with other islands that were never connected. However, although local ecological communities can be explained by historical patterns, the gradients over wide geographical scales are not so easily attributed to time as the major determinant.

The concept that major disturbances drive change undoubtedly carries weight, but closer analysis of tropical rain forests reveals that they too have been subject to disturbances of various sorts and are, in many cases, much changed over geological time. Based on this knowledge Haffer proposed a theory that disturbances in the tropics during the Pleistocene (alternating wet and dry periods giving rise to forest blocks surrounded by grassland) gave rise to refugia in which high levels of speciation took place, ultimately leading to the current high diversity characteristic of the tropics. While this theory is plausible and has been invoked by other authors, tests of the hypothesis have cast doubt over its wide applicability.

Irrespective of the mechanistic bases for variation in species richness over wide latitudinal ranges, time is a significant determinant at a more local scale. Birks (1980) analysed the insect faunas of trees in Britain during the period since the last glaciation 13000 years ago and concluded that there was a significant correlation with the time that a given tree species had been continuously present in Britain during that period. Taking this further, Kennedy and Southwood (1984) included Birks' radiocarbon estimates of time in Britain as a variable in a multiple regression analysis of insects on British trees. They showed that time was a significant predictor, both in its own right and as a co-predictor with log abundance (area). This is logical considering that over time there will be a tendency for well-established tree genera to increase their abundance, despite the recent proliferation in areas arising from commercial afforestation.

**Size**

A consequence of basic structure and longevity is that trees tend to be the largest components of the phylloplane, thus providing a much wider number and range of ecological niches compared with herb and shrub layer plants. Such attributes enable more insect species to colonize the plants without facing competition, both interspecific and intraspecific. Size has many interrelated characteristics that help to explain the presence of both herbivores and their natural enemies.

- A larger physical presence is more easily detected and colonized by invertebrates. This concept, termed 'apparency', reflects a higher visual profile for colonizing adults or passively migrating immature life stages (*e.g.* mean height was a good predictor in models of the distribution of Scolytidae in Finland; Heliovaara and Vaisanen 1995) and a greater production of olfactory cues.
- Complex plant architecture is a consequence of greater size and for trees is particularly linked to the provision of many niches capable of supporting invertebrates with a wide range of feeding strategies (guilds). Although the concept of the guild is not always clearly defined, it is nevertheless a useful means of describing the trophic interactions of invertebrates and trees without having to consider their precise taxonomic status. Thus, seed, bud, leaf, bark and wood feeders can be distinguished, each occupying a different part of the tree and therefore potentially avoiding competition.

- Increased biomass provides greater food resources per individual, thus encouraging growth and breeding success whilst reducing potential intraspecific competition. Price (1992) has reviewed the role of plant resources in insect population dynamics and concludes that, in part, the carrying capacity of an ecosystem depends on plant succession in time and space. Carrying capacity is therefore greater for trees, which generally appear late in succession, compared with herbs or shrubs. Thus biomass is available for longer and in greater quantities on trees compared with other plants.
- The concept of enemy-free space applies to the likelihood of herbivores being overlooked by searching parasitoids or predators. One aspect of size is that there is more opportunity for herbivores to occupy a part of the plant that offers protection against natural enemies (and competition from other herbivores). The complexity and number of niches present on a plant (plant architecture; Lawton 1983) therefore has an effect on enemy-free space. The greater the architectural complexity, the greater the time that natural enemies may have to spend in searching for prey, thus reducing their overall effectiveness and enhancing the survival probability of herbivores.

**Area**

Forests tend to occupy greater areas than other plants. There is evidence to support the view that the greater the area occupied by the host plant, the greater the number of insect species associated with that host. This is called the species-area relationship and was originally linked to the island biogeography theories of species isolation and colonization.

Species-area relationships have been demonstrated for many plants and animals, including both highly mobile and sedentary species. With regard to the numbers of insect species on trees, data on the British insect fauna, one of the best characterized in the world, have been analysed by a number of authors.

Arising from earlier tentative analyses, Kennedy and Southwood (1984) analysed the relationships between insect species and characteristics of tree species in Britain. They concluded that there was a significant relationship between the area occupied by British tree species and the numbers of insect species associated with them. The area occupancy data were based on county and subcounty information of plant records and were thus potentially overestimates of actual areas occupied.

Claridge and Evans (1990) reanalysed the species-area relationships using the same insect data as Kennedy and Southwood but employing actual area information derived from Forestry Commission census data. They concluded that area was not a good descriptor for insect species on deciduous trees but was just significant for conifers.

However, further unpublished analysis of the data broken down into numbers and areas of habitat patches occupied by the tree genera indicated that whereas total area was not a good descriptor, the degree of fragmentation of that area provided a better explanation for the observed insect species numbers. A similar conclusion, concentrating on Lepidoptera that have moved hosts to include conifers, was reached by Fraser and Lawton (1994) in an analysis of new associations on British trees. The overall finding that area provides a broad descriptor of insect diversity remains valid, although this does not explain the mechanisms that lead to these associations.

**Taxonomic Relatedness**

In a given ecosystem, the closer the taxonomic relatedness of host plants, the more similar the numbers of insect species associated with those plants. This has been documented for tree species that have a long history in a particular location and thus provide rich sources of invertebrate species that may subsequently colonize other tree species, especially if they are taxonomically related. A particularly good example of the process in action is the colonization of *Nothofagus,* a member of the Fagaceae newly introduced to Britain, which has been colonized rapidly by insects associated with closely related members of the native Fagaceae.

Geographical isolation as a critical factor in determining invertebrate diversity and implications for international movement of pest organisms. Geographic isolation drives the acceleration of speciation and provides the means to derive principles on the processes involved. In addition, the fact of isolation means that some invertebrates abundant in one region will be absent elsewhere, despite conditions being suitable for their establishment. In such situations, it is pertinent to examine the international plant health implications of geographical isolation and how modern patterns of trade influence the degree of isolation and later colonization by new insect species.

**Geographical Isolation as a Factor in Determining Insect Diversity**

In general, the basic principles apply across ecosystem boundaries and geographical zones extending from the poles to the tropics. In addition, the isolation afforded by relatively small islands provides further evidence of the principles of species colonization over space and time. Great Britain provides a particularly good example, because of its island position in the Atlantic Offshore climatic zone and the fact that it has been subjected to glaciation in recent geological time.

Among the ecological characteristics that arise from this set of circumstances are a relatively depauperate mix of both animal and plant species that have arisen from the combination of losses resulting from the last ice age and the slow rate of recolonization arising from island status. British flora and fauna are therefore impoverished relative to continental Europe. Further influences arise from the preponderance of exotic tree species in commercial forestry in Britain, which although dominated by exotic conifer species also has increasing numbers of exotic broadleaved species.

Expanding on the principle that colonization of tree species is determined by the features, the ranges of numbers of insect species associated with trees in Britain reflect both the area (with provisos) and length of time that trees have been continuously present in Britain. However, there are exceptions, so that Sitka spruce in particular would appear to support far fewer species than predicted on the basis of area, although this probably also reflects the relatively short history of the species and recent expansions in areas planted in Britain.

**Implications for International Movement of Pest Insects**

Trading patterns between countries have had a considerable influence on both the flora and fauna of most of the countries in the world. This reflects deliberate introductions of plants and animals for agricultural, forestry and amenity purposes and also accidental introductions by association with other commodities. In the majority of cases, the introduced species have had a neutral to beneficial effect, reflecting their 'domesticity' within managed crop or animal husbandry systems.

However, there are many exceptions to this generalization and in most cases the results of introduction of new species were not foreseen, either through lack of knowledge of the potential impacts or because the introduction was accidental and undetected. The literature on international biological control is dominated by examples where attempts have been made to rectify the effects of new pest introductions (both animal and plant).

A classic example is the control of the woodwasp *Sirex noctilio* (Hymenoptera: Siricidae) in exotic pine forests in Australia with the introduced nematode *Deladenus siricidicola* (Nematoda: Neo-tylenchidae) isolated from *Sirex* in New Zealand. This nematode has a life cycle that includes a free-living form that feeds on the fungus *Amylostereum areolatum* and a parasitic form that invades *S. noctilio* larvae and subsequently migrates to the reproductive organs of the female wasp, causing sterility. In this situation, all elements of the interaction are exotic to the new location, with little prospect of natural control being realized. ensuring that the resource is conserved to meet the needs of future generations.

FAO works to improve the knowledge on sustainable forest and wildlife management, and supports the development and implementation of appropriate policies and practices to ensure forest and wildlife protection in order to maintain or improve their capacity to produce wood and non-wood products, sustain wildlife populations, conserve biodiversity, safeguard wildlife habitat, mitigate climate change, and protect soils and watersheds.

## What is Biodiversity?

More than 10 million different species of animals, plants, fungi and micro-organisms inhabit the Earth. They and the habitats in which they live represent the world's biological diversity, or biodiversity as it is often called. Humans use at least 40,000 species of plants and animals on a daily basis for food, shelter, clothing and medicinal needs.

Much of the global concern with deforestation focuses on the alarming loss of biodiversity. There is also considerable concern with the poverty of many forest dependent communities. Many poor communities around the world rely on local biodiversity for a range of essential services. These include materials for housing and clothing, food from a range of wildlife species and traditional medicines derived from local plants and animals.

The populations of developed nations also depend on biodiversity for their survival and quality of life. Close to 40 per cent of the pharmaceuticals used in the United States are either based on, or synthesized from, natural compounds found in plants, animals or microorganisms.

The greatest value of biodiversity might still be unknown. Only a fraction of known species has been examined for potential medicinal, agricultural or industrial value. Nor do we fully understand how biodiversity contributes to the well-being of the larger global environment. And we are only just beginning to learn how biodiversity helps communities around the world satisfy their economic, dietary, health and cultural needs.

One thing is certain: the more we learn about biodiversity the more we realize how much the world depends on it. Yet whole species of plants, animals, fungi, and microscopic organisms are being lost at alarming rates.

## Forest Biodiversity

Forests are the most diverse ecosystems on land, because they hold the vast majority of the world's terrestrial species. Some rain forests are among the oldest ecosystems on Earth. Timber, pulpwood, firewood, fodder, meat, cash crops, fish and medicinal plants from the forest provide livelihoods for hundreds of millions of people worldwide. But only a fraction of known species has been examined for potential medicinal, agricultural or industrial value.

## A Continuing Threat

Forest biodiversity is threatened by rapid deforestation, forest fragmentation and degradation, hunting and the arrival of invasive species from other habitats. We are losing 12 million hectares of forest a year, much of it tropical rainforest with its unique and rich biodiversity.

## How can we Protect Biodiversity?

One of the best ways to conserve forest biodiversity is to establish protected forest areas. But these areas must be of a certain size, or consist of a well-designed network of forest areas, to allow the local forest ecosystems to continue operating effectively.

The forest surrounding the protected area must then be carefully managed so that it serves as a buffer zone. These surrounding forests also allow local communities to earn a livelihood without infringing on the protected forest.

There have been numerous efforts aimed at safeguarding the world's biodiversity by protecting species in areas outside their original habitats. For example, seeds of some of the most economically important trees are being conserved in seed centers and gene-banks as a way of protecting their genetic diversity. But a large number of forest species have seed that do not survive storage, and many species of animals and plant-life are hard to protect once removed from their ecosystems.

## CIFOR's Focus

Biodiversity will continue to change and it is impossible to protect everything, so we need to decide which species are critically important and how they can best be maintained. The challenge is to ensure that we do not focus only on the needs of developed nations and the world's economic elite. We must also recognize and examine the priorities of local people who depend on forests, especially in developing nations.

CIFOR is meeting this challenge by studying forest biodiversity and increasing the understanding of this unique resource in ways that accommodate both local and global forest values. It is also researching logging techniques that have a reduced impact on forests and biodiversity while still remaining profitable for logging companies.

# 4

# Seeing the Forest and the Trees: Aquatic Plant Ecology

## HISTORICAL BACKGROUND OF PLANT ECOLOGY

Plant ecology is a branch of the scientific field of ecology which focuses specifically on plant populations. There are a number of applications for plant ecology, ranging from helping people develop low water gardens to studying endangered ecosystems to learn about how they can be protected. Researchers from this field tend to come from an interdisciplinary background which can feature training in a wide variety of scientific pursuits, including plant anatomy, general ecology, biology, and so forth.

The field of plant ecology includes the study of plants and their environment. Rather than just looking at plants in a vacuum, researchers consider how they interact with each other and their environment to create an interconnected system. Plant ecology can include the study of entire ecosystems, such as the rainforest or plateau, or the study of specific areas of interest, like plant populations which manage to survive next to a polluted stream. Plant ecologists also look at animals, soil conditions, and other influences on a plant's environment.

Ecology is a complex and vast field of study which can encompass everything from understanding how natural environments function to how humans interact with the natural world, and how various behaviours can fundamentally alter the natural environment. In plant ecology, people can focus on topics like climate change and its effect on plants, plant evolution, how plants disseminate themselves in nature, symbiotic relationships between plant species, plant diseases, and so forth.

A great deal of field work is involved in plant ecology, as researchers like to see their subjects in nature so that they can learn in context. A single sample of a plant can provide interesting information and data, but actually seeing that plant growing can provide a researcher with a great deal more data. For example, looking at a plant alone, a researcher might not understand why its leaves are shaped the way they are, but when the researcher sees the plant in nature, he or she might realize that the leaves conferred some sort of benefit on the plant or the surrounding environment, ranging from signaling the presence of the plant to pollinators to providing shelter for seedlings so that they can grow up.

Plants make up a vital part of the natural environment, and plant ecologists are well aware of this. In a healthy ecosystem, plants provide food and shelter for animals, secure the soil to prevent erosion, cast shade to create microclimates, conserve water to keep it in the ecosystem instead of allowing it to be lost, and participate in the breakdown and recycling of organic material to keep the ecosystem thriving. Plants are also of critical interest because they produce oxygen, and plants have been heavily implicated in the creation of the Earth's currently oxygen-rich atmosphere.

The relation of organisms and their processes individually to the physical conditions of their internal and external environments is the especial province of the plant physiologist but that of the plant ecologist is to study the integrated effects of the interaction of these physical conditions and the organisms they support, having due regard to the important part played by what we comprehensively term competition. It is the purpose of this article to present some of the salient trends in the study of the causal factors which influence the character and composition of plant communities.

Often in the past it has been found convenient to consider the habitat factors in separate categories, the historic, the climatic, the edaphic and the biotic, but there is an inevitable danger in such abstraction that we underrate the degree to which these interact to affect their influence on plant life. Those influences which have operated in the past to produce the conditions of the present can be classed as historic factors and these, it is manifest, must often have been secular in character.

The replacement of the prevailingly Pteridophytic and Pteridospermous flora of the Coalmeasure Period by the predominantly Gymnospermous vegetation of the succeeding epochs, must have depended upon prolonged changes in the environment, both physical and biotic, as also, no doubt, the transition to the predominantly Phanerogamic vegetation of the present era. The persistence or survival of any one group that marked the facies of plant communities may well reflect, not so much the inherent capacity of the gene complexes involved, to provide organisms of survival value, as their capacity, relative to those of other organisms, to change with sufficient rapidity.

The distinction is important since the one implies an intrinsic lack of physiological plasticity, the other a genetic complex that is merely resistant to adjustment. The one would lead to complete extinction under changing conditions, as with the Pteridosperms, the other to a rare persistence as exemplified by the Royal Fern *(Osmunda)* and the Quill-wort *(Isoetes)* alike survivals of very ancient lineage. Though very remote in the time of their initiation these secular changes are probably still operative in their effects. Indeed the vegetation of the Earth's surface today has, we all recognize, been profoundly modified by the age of steam and the internal combustion engine which alike depend for their source of power on our heritage of fossil fuel from the extinct floras and faunas of the past; their exploitation has accelerated the progressive replacement of natural plant communities by the semi-natural and the quasi-artificial, a transition that the geometrically increasing population of the world is continually augmenting.

If the effects of the remote, and even of the immediate, past are often difficult to evaluate that in no degree denigrates their significance. In our own time man's influence in changing the protective cover of vegetation has disastrously accelerated erosion in many areas. By regulating waterways and creating hard road surfaces he has speeded up drainage and lowered water tables in an ever increasing area, whilst through forest destruction on catchment areas the regulatory influence of the organic sponge has been reduced so that floods are liable to alternate with drought and bring changes far beyond the extent of obvious interference. Anti-erosion measures have recognized the important role that vegetation has to play and the time must one day come when foresters, agriculturists, water engineers and others will recognize a common interest and responsibility in initiating changes in the ecology of the plant cover which they severally exploit.

From the nature of evidences of the past and their often fragmentary character it is only in relation to the most recent events that it is possible to do more than note the contemporaneous occurrence of vegetation changes and physical alterations so that historical ecology is more often descriptive than causal and so outside the scope of this present consideration.

The more obvious climatic influences of wind, precipitation, humidity, temperature, and illumination were, in the earlier ecological studies, considered far too much in isolation from the point of view of the average conditions and with little regard either to their interaction or their indirect effects. Furthermore, it was not till attempts were made to correlate vegetation with the climatic pattern that it was realized how complicated this pattern might be.

## EFFECT OF OROGRAPHIC FACTORS IN MODIFYING THE LOCAL CLIMATE

The effect of orographic factors in modifying the local climate was soon appreciated and the numerous studies in montane areas, such as Switzerland, showed clearly the importance of aspect and of the general level of the mountain massif on altitudinal limits of communities and species. This rapidly led to the recognition of what has been termed cold-air drainage and the existence of frost pockets which could result in temperature inversions and accompanying inversions of altitudinal limits, features that have been closely studied by fruit growers for obvious economic reasons. Our knowledge of the more detailed pattern of climatic features as they affect plant life is, in many areas, inadequate despite the broad correlation between vegetation types and saturation deficits that may obtain in some regions or with length of frost-free periods in others. Moreover our knowledge of the microclimate of the various strata of plant communities and its seasonal variations is generally meagre whilst any concept of the microclimate within the soil is commonly wanting. So, even though we may know the climatic environment of the shoots with some degree of accuracy, we are often quite ignorant as to the significant environment of the root systems.

During prolonged cold spells the depth to which the soil is frozen solid, for example, can vary widely in different plant communities growing on the same type of sub-soil. In one instance bare soil was frozen to a depth 2 to 3½ in., under rough grass to between 1 and 1½ in. and in a hazel copse to only W! in., or not at all, according to the depth of the leaf litter.

## OPERATION OF CLIMATIC FACTORS

The long term operation of climatic factors may clearly limit the capacity of a species to flourish and compete with others or it may even impair or inhibit its power of reproduction. The detailed studies on frost resistance of Apple varieties have, for instance, revealed the diverse times at which temperature-sensitive crises may supervene in the life-history so that if this obtains also in wild populations climatic differentiation in the occurrence of strains as well as species may obtain.

But the elucidation of such relations demands a knowledge, for each species, of these critical periods, such as flower initiation, pollen-tube development, fertilization, etc., and the temperature tolerances of each. The success of any type of plant in the wild state will depend upon its capacity to flourish and reproduce under the local climate which the plant itself in part creates.

The determining conditions are not, however, the average but the extremes to which it is subjected and hence the need for continuous records of the climatic factors within the community itself not forgetting the steep vertical gradients that may obtain between different levels, of the same vegetation unit, and the interference effects of the different species with respect to the penetration of light and rainfall.

The climatic pattern is often conditioned by the structure of the community and the physiognomic character of its constituent members no less than these are themselves determined by the climatic pattern which they can modify and endure. We have referred to the fundamental principle that it is the most severe conditions of habitat, rather than the average, that mainly determine the characteristics of communities. But this, which we may term "the *Principle of Extremes,*" must not be over-simplified and regarded as being in the nature of so many sieves of lethal conditions. Species that will survive a low temperature during a single winter may succumb to two successive winters of no greater severity. Again, it is commonly the extreme conditions, that are not lethal, but sub-lethal, operating from season to season that undermine the vigour of a species and give advantage to the more tolerant vegetation over the less tolerant.

The crude early concepts as to the relationship of plants to low temperature gave place when subjected to experimental tests to the recognition that resistance of a plant or its parts was largely conditioned by its water content and that the capacity of seeds to survive storage in liquid air was only true so long as their water content was negligible.

The degree of toleration of low temperature is well known as a specific character that limits the range of many plants but the complexity of the phenomenon of frost resistance has only been appreciated in the light of recent observations and experiments which have shown that the low temperature toleration of a species varies seasonally and that capacity to endure a certain amount of frost is no guarantee that the same species will not succumb to far less severe conditions if the temperatures be fluctuating.

Similarly with respect to drought tolerance the capacity to endure desiccation is one that is partly inherent, partly dependent upon the time of the year and partly on the manner in which the desiccation comes about. Both with respect to desiccation by drought and by low temperature, the mode and rate of recovery are operative factors in determining survival since rapid expansion of a contracted protoplast, unsupported by the cell wall from which it has shrunk, may involve lethal rupture. A detailed consideration of the general topic of plant hardiness has been furnished by J. Levitt which emphasizes the diverse qualities that render the protoplasm resistant to disorganization but the particular aspect we would here stress is that the precise pattern of climatic change in a particular habitat or partial habitat, often referred to as the microclimate, can alone provide a clue to the climatic factors as they affect the constituents of a natural community.

Thus the seasonal incidence of precipitation if confined to a few thunderstorms in the hottest months may scarcely support a desert type of vegetation whilst the same average rainfall spread evenly over the cooler period may suffice for a plant community which evades the waterless period by shedding its leaves. We must not lose sight of the fact that the precipitation itself may vary to a significant degree in kind as well as in amount.

For example, rain contains chloride corresponding in amount to the proximity of an area to the sea. Dr. Miller of Rothamsted, half a century ago recorded the chloride content as ranging from 218 p.p.m., near the coast, to between 2 and 3.4 p.p.m. fifty miles inland, whilst for sulphur a range of from

0.6 to over ten times that amount has been determined in open country and much higher values in the polluted atmosphere near industrial areas.

## LNOWLEDGE OF CLIMATIC CONDITIONS

Calcium would appear to be present in amounts ranging from one to six parts per million and though such amounts be small their repeated addition to the soil whenever there is a fall of rain does not permit their being ignored. It would, however, seem that variations in the amounts of micronutrients present in rainfall is likely to have but a negligible influence upon plant distribution.

The advances in our knowledge of climatic conditions have emphasized the importance of detailed studies, both spatial and temporal, that take cognizance of all types of humidity and precipitation both above and within the soil. The clue to the presence of vegetation in so-called rainless desert! has been found to rest in the dew formation that even in temperate communities may play an important role.

Quite recently the probable importance of high humidity under exceptional environmental conditions has been further emphasized through the experimental proof by Breazeale*et al.* that plants can absorb water through their leaves from a saturated atmosphere and that this together with contained solutes can exude through the roots into the soil which may thereby attain field capacity.

To descriptive ecology we owe the recognition of a number of major vegetation types that are associated with particular climatic conditions, as for example the occurrence of the deciduous forests in areas where a favourable climate alternates with an unfavourable, due to low temperature or inadequate humidity. Another example is the association of evergreen coniferous forests with areas where the length of the growing season is restricted.

## RELATIONS OF PLANT COMMUNITIES

But now that the broader features of the relations of plant communities with climatic types have come to be recognized the unexplored areas of the world's surface, though likely to provide further examples and interesting variants, are unlikely to contribute materially to our real understanding of the causal factors of distribution unless investigated in considerable detail.

Further progress demands a detailed analysis of climatic and other habitat conditions throughout the seasonal cycle and experimental studies of individual species to determine their tolerances and the critical phases in their development which climatic conditions may promote or inhibit and the temperature and humidity limits these demand.

Other features of the climatic complex that call for more detailed analysis are the responses of individual species to direct and indirect effects of wind action and the influence of radiation of different intensities and wavc lengths. The comprehensive work edited by Professor Duggar constitutes a valuable summary of the position attained at that period, since when the amount of research, especially upon photoperiodism, to which we shall revert, has been both extensive and significant.

In the competition for radiant energy potential heights and spread are morphological features that determine the degree to which one kind of plant or individual can become a sunshade to another but clearly the time factor is here significant and in a population of seedlings, for example, those individuals that germinate first will have an advantage over those germinating later which emphasizes the survival value of what has been termed germination energy.

The more continuous the light screen provided by the foliage of a species the more effective will be its suppression of the plants in its shade though the mechanical vulnerability to wind imposes on leaf area the compromise presented by the compound character of most large leaves, although the rosette-leaves of many biennials which have the support of the soil or other vegetation beneath are immune from this type of injury and are often large and quite entire.

We can, indeed, in many morphological features discern indications of this struggle for light. The dominance of the tree habit in the climax phase where habitat conditions can support forest is but an expression of the fact that in the struggle for light the tallest members tend to survive. The wane of the arboreal vegetation is probably an accompaniment of man's increasing interference with nature making ever-augmented inroads upon woody vegetation so that today, almost throughout the world, herbaceous communities are becoming more conspicuous.

It is, perhaps, due to the importance of the time factor that the more successful species are in general those that have the potentiality of large seed-output and the marked association of average seed size with the shadiness of the plant community that a species normally colonizes is a further emphasis on the importance of the light factor in competition at the juvenile phases of developments. In the competition for nutrients the extent and rate of development of the root system to which we have already referred, is manifestly correlated with the success of the assimilating organs and the mode of spread of a species whether by seed or vegetatively will mainly determine the habitat conditions that it can invade.

From experiments with Barley growing in competition with *Holcus mollis* H. H. Mann and T. W. Barnes concluded that though the root space and supply of plant food was apparently ample for both they reduced the growth of each other and the dominant appeared to be determined by the *time of development* as well as by the density. One could multiply examples but it is clear that temporal factors play an appreciable part in the competition both for light and nutrients and emphasize the need for a much more detailed knowledge than we at present possess concerning the life histories of the chief species contributing to plant communities and the conditions that modify them.

An interesting example of the differential effect of species upon one another was exhibited by some experiments carried out with *Juncus effusus.* A. Lazenby found that the number of seeds of this Rush which germinated and the subsequent growth differed strikingly according to the other species of seeds sown with them. Thus when sown with Bent *(Agrostis tenuis)* and White Clover *(Trifolium repens)* only about one-tenth of the population of Rush plants was obtained as when the accompanying species were Ryegrass *(Lolium perenne)* and White Clover.

It is only natural that the visible, overground, parts of the plant should have received major attention from both morphologists and ecologists though it is perhaps remarkable, having regard to the early recognition of the importance of the root-system, from the nutritional as well as the mechanical aspect, that its nature, extent and mode of development should have so long been neglected. Such neglect is perhaps partially to be explained by the difficulty of exhuming the entire root system, in most soils, as a consequence of which their extent has often been underestimated.

The valuable researches of Weaver and his associates in America have served to call attention to this neglected field of study. From these the considerable labour involved in such studies can be appreciated but also the great diversity and significant distinctions that obtain. Sandy soils afford an especially favourable medium for such investigations since by blowing away the sand from a working face as it dries entire root systems can be excavated without risk of losing the finer ramifications and by appropriate means these can be supported in their original positions as they emerge from the matrix. The present writer's own studies on the root systems of plants characteristic of various phases of dune development have shown how varied these can be not merely between different species but also between different individuals of the same species when growing in conditions of diverse water supply and aeration. Root studies have emphasized the distinction between those that exhibit a restricted but richly branched system, which exploit a relatively small volume of soil intensively, and those in which the root system is extensive but the large soil volume is much less completely interpenetrated.

Modern studies of economic plants have shown what marked differences in this respect may obtain between cultivars of the same species as, for example, the root systems of the "Wealthy" and "Doucin" apples or the various East Malling Apple stocks, and there is no reason to suppose that similar differences both interspecific and intraspecific, do not occur with respect to the root systems of wild plants and that they play an important role in their competitive equipment. The recognition of the fundamental importance of a number of trace elements and their frequent presence in inadequate amounts, or in more or less non-available condition, has brought into prominence the great significance that may well attach to the type and degree of root exploitation.

Furthermore it has been shown that the availability of some trace elements, such as zinc, in the early stage of development of a species, can be decisive for the subsequent healthy growth and its early supply will differentiate between failure and success. The importance therefore of the *rate* of development of the root systems of different kinds of plants in what we may perhaps be permitted to describe as their race for the necessary nutrients which are in short supply, scarcely requires emphasis. The great success of the grasses may be in no small degree due to the extent and rapidity of their root development. Weaver showed that the latter is frequently more than 12 mm in a day whilst the root of a maize plant may extend more than 50-60 mm in the same period.

With uniform genetic material the extent of the root system is very susceptible to the influence of water soluble substances, for example the calcium ion is known markedly to affect root growth. The extent of the root surface in the presence of calcium was several times that in its absence and the volume of soil that can be exploited is probably proportionately increased. The extent of the root system is also affected by the soil texture and indeed one aspect of cultivation is that it reduces the resistance that the growing root has to overcome in its penetration of the soil. Again, optimum moisture content and above all good aeration promote not merely the extent of the root system but in particular the development of root hairs.

One may note that root hairs can augment the absorbtive surface sevenfold, so that factors affecting their production and growth may be of great significance. All these conditions affecting the size and efficiency of the absorbing structures stress the importance of the morphological-physiological relations and one may expect that future ecological studies will tend to integrate still further the two angles of approach, but if this is to be achieved it is essential that the physiologist should be apprehensive of the extent to which field competition is likely to modify the laboratory indications and the ecologist must be alert to appreciate the field significance of the controlled experiments which, however oversimplified they may be from the point of view of their translation to the plant community, can rarely fail to make their contribution to our understanding of the more complex conditions.

In 1929 I called attention to the striking reduction in the extent of the root system of a species when subjected to the competitive influence of the root systems of other plants. Eight years later Pavlychenko furnished remarkable quantitative data respecting the magnitude of this effect as exhibited by cereals in competition with weeds and with one another. The importance of the competition factors in natural communities cannot be over-emphasized and such may well be operative from the earliest phases of plant development since seedling mortality, which is very high, appears to be influenced not only by the physical conditions but by the proportions in which the seedlings of different species are present.

Statistical studies have confirmed what general observation had suggested that individual plants are not normally dispersed in a random manner. This was for instance clearly brought out in regard to grassland communities by G. E. Blackman, in 1935, but it yet remains to be ascertained to what extent such lack of uniformity in dispersion reflects the lack of uniformity in the habitat conditions. For example, the marked tendency towards aggregation of the individuals of the herbaceous species of *Salicornia* is clearly to be expected from the crozier-like hairs upon the testa which anchor the seeds to suitable surfaces, such as the wefts of salt-marsh algae, and to one another, when they are dispersed by the rising and falling tides.

So, too, the classical examples of the hygroscopic arms of the dissected spore-coats of *Equisetum* which ensure some measure of aggregation of the unisexual gametophytes. These and other similar morphological determinants may clearly be important for survival but unrelated to any permanent feature of the habitats concerned. We still do not know why it is that the annual vegetation of a desert, superficially uniform, may present a Joseph's coat with large areas each dominated by a different species of *Helipterum.*

Are such phenomena visual expressions of vagaries of dispersal, or evidences of survival from selection accompanying real though obscure habitat conditions? Attention has more than once been drawn to the phenomenon of certain species that occur with great abundance in an area and produce copious seed only to disappear for a long period subsequently or, if not completely absent, they are rare. A striking example that has been observed both in this country and on the Continent is afforded by the Yellow-wort *(Blackstonia perfoliata),* which is very pertinent to our present context because its root system is of comparatively small extent.

The output of its small, almost dust-like, seeds which exhibit a high per centage germination is large. So it is not unreasonable to assume that these are in fact broadcast over the entire area. Hence, one may ask, is it that such species require some nutrient condition, perhaps a trace element in short supply and that this becomes depleted? The discontinuity of the densest population of *Blackstonia* would not warrant suggesting that its roots completely exploit the entire surface soil but it may well be that, from a vast population of seedlings, the mature plants we see are in fact survivors where the requisite nutritional conditions obtain and that the discontinuity of occurrence does indeed reflect the heterogeneity in this respect of a habitat otherwise apparently uniform.

## ROLE OF VEGETATION IN THE PROCESSES OF PEDOGENESIS

The role of vegetation in the processes of pedogenesis has been studied under conditions where the distribution of the phases of plant development in space corresponds to their sequence in time and thus provides an unequivocal chronological succession over a prolonged period with respect alike to soils and organisms. For sand dunes it has been shown that the soil development is dependent upon a succession of species that incorporate organic material beneath the surface as well as providing deposition upon it and augment water retention and stability so that less and less specialized types can effect colonization ultimately ousting the pioneer specialists.

The original inhospitable soil conditions with an open community of perennial and annual species, characterized by features that enable them to endure the rigourous conditions, is succeeded by conditions favourable to an increasing number of species, far more diverse in their biological equipment, that ultimately constitute a closed community and this in turn may finally give place to scrub and woodland occupying soil that is now favourable to a great diversity of plants but where in fact the number of species has again diminished owing to the dominance of trees and shrubs.

The initial edaphic specialization has thus, in the course of perhaps a century, given place to a biotic specialization that in its own way may be almost equally severe. The studies of R. L. Crocker and his associates on the recessional moraines deal with a comparable sequence from an open community of pioneers that in the course of some 120 years has developed to spruce forest and, as in the dune succession, there is an increasing organic content and transition from an initially alkaline soil to one of appreciable acidity with an augmenting gradient in the soil profile.

The recognition of the major role in water retention by the soil played by the organic material gives added significance to its vertical distribution and it thus came to be realized that, in the naturally stratified soil the surface layer might be of sufficient thickness and its organic content so high that it constituted a sponge which could starve the lower layers of water except perhaps in the heaviest rainfall. Thus is indicated one of the ways in which during the course of years a tall deeply rooted layer of dominant species continually adding its quota to the superficial organic accumulations can gradually contribute to its own destruction and an acidophile shallow-rooted community replaces the erstwhile dominants.

Each species has its own complex of conditions in which it can best develop. If we grow any one in conditions that vary with respect to one factor, such for instance as soil reaction, we find there is a range over which the plants exhibit maximum growth on either side of which increasing acidity or increasing alkalinity is accompanied by diminishing vigour. In the absence of competition the range of toleration is often wide but in the presence of other species this may become very restricted.

This fact emphasizes the fundamental principle of the interaction of habitat factors. For whether it be soil reaction or any other there is usually no absolute optimum but a relativity that is the more complex because any single one may operate in so many different ways. Soil acidity, for example, may be beneficial as enabling certain nutrient ions to be readily available but it may also be directly harmful by reason of toxicity of the hydrogen ions themselves, or indirectly through excess of ions of alumina or manganese in the soil solution, or again it may adversely affect beneficial soil organisms such as the nitrogen fixing bacteria and encourage harmful ones such as *Plasmodiophora brassicae.*

Above all, adverse or beneficial factors cannot be considered in isolation since plants grow where they must and not where they will so that what, in the laboratory experiment, may seem harmful appears in the field as a benefit because the species concerned is more tolerant of this condition than its normal associates. For instance *Viscaria alpina* can apparently tolerate a copper content in the soil toxic to most species. Again what favours vegetative growth is often detrimental to seed production and which of these is the more important may vary from time to time in one and the same plant community. Experiments have repeatedly shown that the optimum conditions with respect to various factors alter with the stage of development of the individual and also its susceptibility.

"The field for the problem and the laboratory for the solution" is only very partially true for ecological studies, for laboratory findings must be transmuted into the far more complex context of so many other variables that experience alone can guide as to what may be applicable and to what degree. But field studies, both observational and experimental, should serve as the means for testing hypotheses based upon what the plant physiologist has ascertained. Ecology is essentially applied plant physiology with the competition factor as an important modifying influence in a physical environment that represents the interaction of climatic, edaphic, and biotic factors that are themselves abstractions from a sequence of continual change.

The emphasis on physiognomy that has formed the basis for some systems of classification of plant communities can be regarded as an attempt to utilize the plants themselves as integrators of the physical conditions and the concept of the so-called indicator species is but another aspect of the same theme. When extreme conditions are under consideration the correlation of physiognomy with habitat conditions is sometimes surprisingly high as is well exhibited, for example, on some of the sandy and areas in South-West Australia, where communities are encountered consisting of species of the most diverse genera and families yet resembling one another so closely in their general physiognomy as to be difficult to distinguish except when in flower or fruit.

In less differentiated conditions, however, as for instance in the temperate woodland, the same stratum, where there is relative uniformity of environment, often presents a great range of morphological diversity that suggests either that physiognomy is an unsatisfactory guide or that our knowledge must be greatly increased by experiments, in the field of environmental morphogenesis, before we are in a position to assess what morphological characteristics are significant and what are irrelevant.

## STRATIFICATION AND SEASONAL DEVELOPMENT OF AERIAL SHOOTS

The stratification and seasonal development of aerial shoots, to which attention was drawn by Kerner Von Marilau nearly a century ago, and of the subterranean organs, which the late Dr. Woodhead was one of the first to investigate, both affect the competitive efficiency of the species whilst the periodicity of the important phases of the life history and their modifications by the environment are no less significant though too little investigated.

These latter, for economic reasons, are better known with respect to horticultural plants. The gardener is familiar with the fact that most plants are best moved in the dormant state, yet most species of *Helleborus* thrive best if transplanted when in full flower, a difference probably related to the phases of root development. Or again, careful investigations of bulbous species have shown how temperature fluctuations can affect differently diverse cultivars of the same species apparently by reason of their individual rhythms in the production of flower initials.

What is true of cultivated plants is probably equally true of wild species and emphasizes for us that the understanding of the communal life of plants is dependent upon a knowledge of the idiosyncrasies of its chief constituents. The structure and chemical characteristics of the soil as they affected plant life gained no small impetus from the work of Lawes and Gilbert, initiated in the early forties of last century, but the classical work of E. J. Russell on *Soil Conditions and Plant Growth,* which first appeared in 1912 and which passed through eight editions and has been translated into various languages, probably did more than any other single publication to bring before the minds of botanists the complexity as well as the importance of edaphic conditions upon vegetation.

The soils that the agriculturist had studied were, however, to a very considerable degree physically homogenized by the processes of cultivation and chemically altered by their manurial treatment. It was not until ecologists began to pay attention to the organization that is exhibited by natural soils that the degree of applicability of the findings of the agriculturists could begin to be assessed. One of the earliest recognitions of the structure of natural soils and its relation to the vegetation which they bear was the study by Gesser and Siegrist of the profiles in the soils of the Aare communities in Switzerland. In this country the present writer called attention to the striking vertical gradients with respect to organic content, water content, soil reaction, and other features exhibited by woodland and sand-dune soils which provide a gradient of conditions for the root system.

These vertical changes not only contribute to the establishment and maintenance of the complementary communities of species but, as soils derived from a calcareous substratum strikingly show, can provide one and the same root system with a diversity of conditions favourable for the easy acquisition of the whole range of nutrient requirement from an acid surface layer from which the ions of iron and manganese may be readily obtained to an alkaline subsoil from which molybdenum is readily available.

## SYSTEMS OF SOIL CLASSIFICATION

The systems of soil classification developed particularly in relation to highly differentiated climatic conditions, as for example by K. Glinka, soon led to the recognition of climatic soil types but also the appreciation that under less extreme conditions, such as may obtain in western Europe and especially in Britain, the climatic conditions can cease to be the main factor in soil genesis and the nature of the substrate from which the soil is derived becomes of great significance.

In Britain the criss-cross, of changing climatic chara-cteristics from East to West and North to South, superimposed upon the rapidly changing geological strata from the more recent towards the South and East to the older in the West and North, has produced in the British Isles a peculiarly intricate mosaic of soil conditions that has been still further complicated by the glaciations and the post-glacial loess that now appears to have been even more widespread than had hitherto been envisaged as shown by the investigations of Perrin, Coombe and others.

These various soils may enhance or ameliorate the effect of present-day local climate and this has enabled a greater diversity of immigrant species to persist in British plant communities than might otherwise be able to survive. For a long time the significance of soil aeration was ignored and Clements in 1921 did a service to this aspect of ecology by bringing together much of the relevant literature. Recent work would seem to emphasize its importance not merely by reason of its influence on the extent of the root development but also because the actual intake of the nutrients appears to be directly related to efficient aeration of the adsorbing cells that can provide the energy for intake against an adverse gradient.

From the data furnished by Hopkins and his associates it would appear that the intake, of plant nutrients generally, decreases with the reduction of the oxygen content of the atmosphere that surrounds the root-system. The effect upon the intake of the trace elements appears to be less than for the major nutrients whilst sodium intake would seem actually to augment with diminished oxygen supply, a feature that may be of considerable import for vegetation growing on soils liable to flooding by sea-water.

Both the oxygen content of the soil-atmosphere and its carbon-dioxide content are important and depend upon the activity of soil organisms and the freedom of diffusion with the air above, which latter is facilitated by volumetric changes due to temperature fluctuations. So it is that under turf or woodland litter high carbon-dioxide concentrations may obtain, in illustration of which we may note that Russell and Appleyard recorded over 3 per cent under grassland and Harley and Brierley under beech-leaf litter from three to eleven times the amount in the atmosphere above.

A high organic content is often associated with a large microbial population so that the production of carbon dioxide is copious and the surface litter tends to retard diffusion as well as insulating the surface soil from rapid temperature fluctuations. In assessing the climatic and edaphic factors as they affect plant life we have in the past been perhaps too apt to regard climate and soil conditions as separate abstractions so that a true soil ecology has been distorted by the partialities that are only valid in isolation. The investigations of plant physiologists during recent decades have contributed to our appreciation of the direct influence of environment upon vegetation, notably in relation to two types of factors, the climatic and the edaphic.

It is now over thirty years since Garner and Allard published the results of their classical experiments demonstrating the influence of the relative duration of daylight and darkness upon growth and reproduction. Since that time the vast volume of literature on the photoperiodic response has revealed that, for a large number of species of flowering plants, flower formation is dependent either on the stimulus of a short period of darkness and a long day or the stimulus of a longer period of darkness and a shorter day. Variations in the relative duration of day and night could moreover retard or inhibit reproduction by seed and also influence the degree and character of vegetative growth.

For annual species whose growth rhythm is adjusted to such stimuli the latitudinal changes in photoperiodism may clearly operate as a limiting factor to their distribution, but for spreading perennials the substitution of vegetative multiplication for the production of fruit may well extend their geographical range. *Mercurialis perennis,* for example, so abundant in our basic woodlands, achieves success by means of its rhizomatous aggression although the reproductive capacity from seed towards its northern limits is negligible.

The specialization, which response to such stimuli implies, may therefore be advantageous, or the reverse, when the species concerned extends beyond the climatic conditions of its original home. It is significant therefore to note that some species have not developed any response relation to photoperiodic stimulation, as a time keeper of their growth rhythms, and that these include some of our commonest weeds such as the Chickweed *(Stellaria media),* the Shepherd's Purse *(Capsella spp.),* and Poa annua which flower and fruit at all seasons when temperature conditions permit and are well-nigh cosmopolitan in their distribution.

For these and other such species the lack of photoperiodic response is manifestly a factor in their success and determines in part the role they play in vegetation. It is also to be noted that photoperiodic response can be markedly influenced by temperature and humidity whilst within one and the same species long-day and short-day strains can be evolved. Probably the chief contribution of physiology to an understanding of the habitat's direct influence in recent years was the discovery of what have been termed the micronutrients or trace elements. That extremely small quantities of certain elements, mostly only requisite in parts per million, were nevertheless essential for plant life is due to their role in forming catalytic agents of various metabolic activities. Such micronutrients now regarded as generally necessary for the higher plants are boron, copper, zinc, manganese, and molybdenum.

But in addition to these, various other elements would appear to be at least beneficial, if not actually essential, to certain plants, as for example sodium and iodine, whilst though there is some experimental evidence for stimulation of growth by minute traces of, for example, titanium and arsenic, it seems unlikely that these play any significant part in determining the composition of plant communities or the degree of dominance of species. It should, however, always be borne in mind that owing to the "compound interest law" minute initial advantages may lead to significant final effects.

## QUANTITATIVE ESTIMATION OF AMOUNTS

The quantitative estimation of amounts of such diminitude as that in which the trace elements occur, with an accuracy of within 10 per cent or less, is one of the triumphs of modern spectroscopic technology and these determinations have shown that deficiencies may often occur either through their inadequacy in the soil, through their absence, or through being rendered non-available.

The last condition is exemplified in respect to manganese, zinc, and boron in markedly alkaline soils whilst copper deficiencies or manganese deficiencies can be brought about by microbiological activity. The observation that certain characteristic appearances are associated with deficiencies of each of the various nutrients, both macro- and micro- alike, made practicable the survey of such occurrences over a far wider area than would have been possible if analyses alone could have sufficed. The outcome has been the recognition that the availability of most of the essential elements for the nutrition of the higher plants is very variable. In natural soils rainwater charged with $CO_2$ readily brings the calcium bicarbonate into solution and by base exchange with potassium and magnesium these also are readily leached from the soil so that deficiencies of all three are by no means uncommon.

Calcium is never wholly absent from soils that can support higher plants since it is an essential constituent of their cell walls but the great range of concentration, from the so-called non-calcareous soils, devoid of free carbonates, to the highly calcareous soils of downs and dunes, is accompanied by a corresponding change in the character of the vegetation which may involve the replacement of a calcifuge by a calcicole species occupying a similar niche or the change from acid moorland to calcareous grassland. A recent review of the role of magnesium as a plant nutrient shows that many natural soils in Europe may contain very small amounts, sometimes less than 0.01 per cent and magnesium deficiency is, in fact, frequent. The large areas in South Australia that have been rendered fertile, by amending deficiencies of molybdenum, zinc, and copper, is sufficient evidence for the ecological significance of the distribution and availability of trace elements. Though this is admittedly an extreme example the areas where deficiencies of one type or another have been recognized are continually being added to.

It is of considerable import that evidence has been furnished that some species are much more susceptible to such deficiencies than others whilst some plants can accumulate trace elements in a marked degree so that, when these are essential nutrients, such capacity must be an important asset in competition. On seleniferous soils grass herbage may be quite innocuous to stock whilst species of *Astragalus* if present can accumulate selenium to such an extent as to render the herbage lethal.

So, too, clovers will accumulate molybdenum and the Sun Spurge, *Euphorbia helioscopia,* is known to be able to accumulate boron, facts which suggest that competition for the micronutrients may, because they are frequently marginal in concentration, be more severe than for the macronutrients which are more commonly considered from this standpoint. Of these latter, deficiency of nitrogen is common in acid soils and those containing little organic material, whilst phosphate deficiency also characterizes acid soils in areas of high rainfall but especially when these are rich in organic matter since phosphorous in organic combination is even less readily available to the plant than from inorganic compounds such as iron-phosphate. Potassium deficiency is common on chalk, sandy soils, and peat.

Although much is known as to the respective demands that crop plants make upon the various nutrients and of their susceptibility to toxic concentrations of manganese or aluminium, our knowledge of wild species in these respects and of their powers of accumulation of nutrient ions is as yet fragmentary although these might well furnish the clue to many apparent idiosyncrasies of presence and absence in communities.

That the precise form in which an element is present may be far more significant than its amount is manifest from the efficiency of chelated compounds of iron and manganese for relieving deficiencies of these elements where other compounds are wholly ineffectual. For the ecologist, as for the agricultural chemist, the problem is not the accurate determination of the total amounts of the various nutrients present but the estimation of the proportion of each which is in a form that the plant can assimilate.

A recent investigation of New Zealand soils by D. E. Hogg, with respect to potassium, is illuminating in this connection. Utilizing a number of standard methods of assessment he found that determinations of the exchangeable potassium corresponded closely to the response made by plants when potassium was supplied but only in the regions of high rainfall. In the areas of low rainfall assessment based upon nitric acid extraction showed a closer relationship with crop response.

It is manifest that the varying availability of the macro- and micro-nutrients may have a direct effect upon the distribution of species and communities but the possibilities of indirect effects must not be ignored. We have already referred to the differential absorption of such potentially lethal elements as selenium and molybdenum which can be economically serious in relation to agriculture, but under natural conditions such differences might materially influence the biotic pressure exerted by the grazing animal.

So obvious is it that a plant community is a social aggregate of its constituent species that one is apt to forget its implication, that all these must have similar environmental preferences or tolerances, except in so far as the presence of one or more species ameliorate, or exacerbate, the conditions for others. The obvious manifestation of this is the creation of shelter by taller species but there may be many other influences more subtle, as for instance the capacity of some species to tolerate and accumulate ions of alumina and manganese which may conceivably render conditions acceptable to other plants more susceptible to their toxic action.

It may be urged with some justice that many of the features we have mentioned as affecting distribution are mostly as yet only known to operate locally and determine the success of individual species rather than communities. That individual species can bring about vegetation changes of a widespread character is, however, witnessed by the disastrous extension of *Opuntia inermis* in Queensland or the spectacular spread of RosebayWillow-herb in Britain. It would appear to be the initial establishment and the build up of the population of a species to the requisite "infection pressure" that presents the major obstacles to marked extension and it is not least in this context that one should perhaps take particular cognizance of such discriminating causes.

So far we have made no reference to one class of biotic influences, namely the effects of animal life upon the nature of the plant community. No doubt in some degree it is true that the vigour of growth of vegetation and its rate of reproduction is adjusted to the hazards of the predators upon it, but it has been shown that Herbert Spencer's facile generalization is invalid and even were it true the imposed burden for survival is a serious handicap in competition. Recently C. B. Huffaker has expressed the view that "changes in plant populations and insect populations are often reciprocally related phenomena". The profound effect of a high rabbit population in Australia and in Britain and its drastic diminution by myxomatosis has recently brought into prominence the differential influence that an animal population can impose. Only in recent years have population studies shown what high numbers vegetable feeders such as field-voles can attain and whilst the importance of animal life, both large and small, is doubtless profound, yet the assessment of its differential effects on plant communities still remains largely the subject-matter for future research.

Competition between species and its severity in a plant community is dependent upon the requirements of the constituents both as to what demands each makes upon the habitat and the degree of intensity of these demands with respect to time. All flowering plants require water and mineral nutrients and most require light but the claims made upon these may vary greatly with the species concerned.

Competition ensues only when the supply is inadequate to meet the full requirements of all, a condition which may be attained with respect to light in open communities and with respect to water in aquatic ones. Competition for light we have already briefly referred to. It is largely dependent upon potential height and leaf area on the one hand and on the other upon the specific compensation point. Competition for water and nutrients is clearly dependent upon the differing capacity of a species to obtain these and to develop a root system, in the presence of others, adequate to its demands with sufficient rapidity and this can only be accomplished if, *inter alia,* the requisite photosynthetic products are available to effect this growth.

In assessing the adequacy of supplies of essentials it is necessary to distinguish, as in the human counterpart, between mere subsistence and what is requisite for vigourous growth. It has been asserted that there is no competition between the herbaceous layer of a forest and the dominant trees but, as we have pointed out, the former may deprive the deeper-rooted trees of their water supply although in general it is between the species of the same stratum that competition is most severe. Competition is perhaps most acute in the early stages of growth when mortality is highest.

Between individuals of the same species an equilibrium may be established so that over a wide range of population density the total productivity is very similar as has been shown in numerous crop spacing experiments. Between different species the reactions vary greatly. Some, which are closely allied and make similar demands upon the habitat (*e.g. Cerastium semidecandrum and C. tetrandrum*) may attain a measure of equilibrium but with open communities priority of establishment is significant and hence the mode and times of germination and the processes of dispersal play their part.

In perennials the mode of growth and rate of potential spread can play significant roles. Plant succession can be thought of as in large part a competitive drift in which at each phase until the climax the constituent species render the habitat more favourable to their successors than to themselves. From this aspect the few species that appear to be able to establish an almost static condition, gain a special significance. The maritime communities dominated by *Halimione portulacoides and Spartina townsendii* may be cited as possible examples though their seeming stability may be but an illusion due to the slowness of their change. The importance of the competition aspect of succession is that it emphasizes the slow rate at which competitive influences may operate, masking the almost imperceptible decline or augmentation of individual species despite the vagaries of seasonal fluctuations.

One of the desiderata is for more actual data as to the time scale involved. We know, for instance, that weeds of arable land that reverts may survive for perhaps half a century, that some pioneer dune species may persist nearly a century, but the rate of a decline under various degrees of competitive pressure for the characteristic perennial species of communities could materially assist in estimating the relative importance of reproduction, vegetative spread, and life span as factors in aggressive capacity.

From this brief survey it is apparent that future progress demands a far more detailed analysis of the habitat factors alike in space and time and parallel with this a more meticulous assessment of the constituent organisms amongst which the importance of sub-specific taxa may far transcend the specific aggregates in ecological significance, alike with respect to their biological behaviour and their physiological attributes.

## THE ELEMENTS OF COMPLETE PLANT NUTRITION

Complete nutrition results in superior plant growth. Why choose anything less for your plants? There are 20 elements necessary for optimum plant growth. Air and water supply carbon, hydrogen and oxygen. Macronutrients are required by plants in large amounts. Micronutrients are required in smaller amounts. Eliminate any of these elements,and plants will display abnormal growth, deficiencies or may not reproduce. The following is a brief guide to the role played by each of these essential nutrient elements.

### MACRONUTRIENTS

- *(N)-Nitrogen*: Component of proteins, hormones, chlorophyll, vitamins and enzymes. Promotes stem and leaf growth. The ammoniacal and nitrate forms are used directly by plants for stem and leaf growth. The urea form of nitrogen must be broken down by soil borne microorganisms or urease before it can be utilized by the plant. Urea can cause leaf tip and root burn. Deficiency (Def.): reduced yields, yellowing of leaves, stunted growth. Excess nitrogen can delay fruiting and flowering.
- *(P)-Phosphorus*: Seed germination, photosynthesis, protein formation, overall growth and metabolism, flower and fruit formation. Def.: purple stems and leaves, retarded growth and maturity, poor flowering and fruiting. Large amounts without zinc cause zinc deficiency. Low pH (<4) ties up phosphates in organic soils.

- *(K)-Potassium*: Formation of sugars, carbohydrates, proteins, cell division. Adjusts water balance; improves stem rigidity and cold hardiness; enhances flavour, colour and oil content of fruits; important for leafy crops. Def.: spotted, curled or burned look to leaves; lower yields.
- *(Ca)-Calcium*: Activates enzymes; structural part of cell walls; influences water movement, cell growth and division Required for uptake of nitrogen and other minerals. Leached from soil by watering. Immobile-requires a constant supply for growth. Def.: stunting of new growth in stems, flowers, roots; black spots on leaves and fruit; yellow leaf margins.
- *(Mg)-Magnesium*: Critical component of chlorophyll; needed for functioning of enzymes for carbohydrates, sugars and fats; fruit and nut formation; germination of seeds.
- Def.: yellowing between veins of older leaves; chlorosis; leaf droop. Leached by watering. Foliar spray to correct deficiencies.
- *(S)-Sulfur*: Component of amino acids, proteins, vitamins, enzymes. Essential for chlorophyll. Imparts flavour to many vegetables. Def.: light green leaves. Water supply may contain sulfur. Leached by watering.

**MICRONUTRIENTS**

- *(B)-Boron*: Affects at least 16 functions: flowering, pollen germination, fruiting, cell division, water relationships, movement of hormones, cell wall formation, membrane integrity, calcium uptake, movement of sugars. Immobile; easily leached. Def.: terminal bud die back causes rosette of thick, curled, brittle leaves or brown, discoloured, cracked fruits, tubers and roots.
- *(Cl)-Chlorine*: Involved in osmosis (movement of water or solutes in cells), ionic balance necessary to take up mineral elements and photosynthesis. Def.: wilting, stubby roots, yellowing, bronzing. Scents in some plants may be decreased. Leached by watering.
- *(Co)-Cobalt*: Required by nitrogen fixing bacteria; formation of B12 vitamin; formation of DNA. Will extend life of cut flowers such as roses. Def.: may result in nitrogen deficiency.
- *(Cu)-Copper*: Necessary for nitrogen metabolism; component of enzymes - may be part of enzyme systems that use carbohydrates and proteins. Bound tightly in organic matter. May be deficient in highly organic soils. Not readily lost from soil but may be unavailable. Def.: die back of shoot tips; terminal leaves develop brown spots. Excess is toxic.

- *(Fe)-Iron*: Enzyme functions; catalyst for synthesis of chlorophyll; essential for new growth. Def.: pale leaves, yellowing of leaves and veins. Leached by water and held in lower parts of soil. High pH soils may have iron present but unavailable to plants.
- *(Mn)-Manganese*: Enzyme activity for photosynthesis, respiration and nitrogen metabolism. Def.: young leaves are pale with green veins similar to iron deficiency; advanced stages-leaves are white and drop; brown, black or gray spots may appear next to veins. Plants in neutral or alkaline soils often show def. Acid soils may increase uptake causing toxicity.
- *(Mo)-Molybdenum*: Structural part of enzymes that reduce nitrates to ammonia for amino acid development essential to protein formation; required by nitrogen fixing bacteria. Def.: pale leaves with rolled, cupped margins. Seeds may not form. Nitrogen deficiency may occur if plants are lacking Mo.
- *(Ni)-Nickel*: Recently recognized as essential. Required for the urease enzyme to break down urea into usable nitrogen and for iron uptake. Seeds require nickel to germinate.
- *(Na)-Sodium*: Improves nitrogen metabolism in many plants, involved in osmotic (water movement) and ionic balance in plants. Def.: yellowing of leaves and leaf tip burn; may inhibit flower formation.
- *(Si)-Silicon*: Component of cell walls; enhances resistance to sucking insects and fungi.Foliar sprays reduce populations of aphids on some plants. Enhances leaf presentation; improves heat, drought and cold tolerance; improves photosynthesis; extends bloom life. Def.: wilting, poor fruit and flower set, increased susceptibility to insects and disease. Disease resistance is enhanced by regular foliar feeding.
- *(Zn)-Zinc*: Functional part of enzymes including auxin (growth hormone) synthesis, carbohydrate metabolism, protein synthesis, stem growth. Def.: mottled leaves, irregular yellow areas. Zinc deficiency leads to iron deficiency. Occurs in eroded soils; least available at pH of 5.5-7.0. Lower pH can cause availability to the point of toxicity.

## NUTRIENT BALANCE AND NUTRIENT DEFICIENCIES

Plants (like people) need a 'balanced diet'. They need all 13 nutrients to remain healthy. If one is missing, the plant will not grow well. Poor plant nutrition causes plants to grow slowly in the nursery and in the field, and to be more susceptible to diseases. Many people confuse the symptoms of nutrient deficiencies with those of too much or not enough shade or water.

In fact, all three factors, shade, water and nutrients affect plant growth, and interact to produce healthy plants. A plant that grows in full light with abundant moisture and receives all the 13 nutrients will grow fast and have a dark green colour in its leaves. A plant that grows slowly in the shade may also have dark green leaves, but when exposed gradually to the sun, the leaves may turn yellow.

This does not mean that plants do not like full sun—it might indicate a nutrient deficiency which did not show up in the shade because the plant did not have enough light to stimulate fast growth. Together, water, shade and nutrients must be monitored and adjusted to produce quality seedlings. It takes practice to learn the signs identifying a missing nutrient or nutrients, but you can learn to do so, and some of the signs are common to many plants. A *good nursery practice* is to carefully monitor the leavcs of your plants for signs of nutrient deficiency, and correct them with a better substrate or with fertilizer.

## INORGANIC FERTILIZERS

Inorganic fertilizers are mined from the soil, or produced during complicated chemical reactions. A *good nursery practice* is to read the fertilizer labels. This allows you to apply what the plants need without wasting nursery resources. Fertilizers contain only plant nutrients; they are not used to combat plant diseases or insects. Inorganic fertilizers do not improve the substrate physical properties, whereas organic material such as compost does. Inorganic fertilizers are also expensive and not always available in the stores. Nursery managers should carefully consider the cost and benefit of buying these products.

Granular fertilizers are commonly given names like "17-17-17", or "10-30-10". What do the numbers mean? They represent the percentages of nitrogen (N), phosphorus (P), and potassium (K) in the fertilizer—17 per cent N, 17 per cent P, 17 per cent K. In this case, 51 per cent of the mixture is made up of N-P-K, and the rest is inactive material used to help spread the fertilizer evenly. Urea contains only N, and is labelled as 46-0-0. Urea is very strong and can easily burn the plants if too much is applied.

## GRANULAR FERTILIZERS

Granular fertilizers can be mixed into the substrate or into the irrigation water, or be applied to older plants on the soil surface. It is better to mix the fertilizer directly into the substrate before planting the seed because the roots can avoid or seek the fertilizer as they need it. Use only small quantities such as 2 or 4 grams (1/2 teaspoon) per 1 kg of soil. It is better to add too little then too much.

You need to experiment with different levels. Plants should respond within two weeks. When dissolving fertilizer in warm water, carefully note whether it is thoroughly dissolved. If not, it is probably the phosphorus that remains. It may be better to apply fertilizer in granular form if it does not dissolve thoroughly. Apply liquid fertilizer to the soil, not to the leaves which are easily burnt if fertilizer remains on them. Be extremely cautious when applying fertilizer to young plants.

## FOLIAR FERTILIZERS

Foliar fertilizers are used in order to get the nutrients to the plants quickly. They are specially formulated to apply directly on the leaves. Foliar fertilizers are absorbed by the leaves, not by the roots. When plants are acutely deficient in nutrients, foliar fertilizers often help 'green them up'. Frequently, foliar fertilizers only contain the micronutrients, since it is assumed that the macronutrients are available in the substrate. However, some such as 'GroGreen'® contain both micronutrients and 20-30-10 of N-P-K. Often an adhering agent such as 'Da-Plus' is used to help the fertilizer stay on the leaves so that it is not washed off in the rain. Because foliar fertilizers are expensive, and they may not encourage strong root growth, they should not be used as a long-term solution for plant nutrients.

## COMMON NUTRIENT DEFICIENCY SYMPTOMS OF MACRONUTRIENTS

- *Nitrogen*: This is a mobile nutrient, which means that when nitrogen is deficient, plants move it from the older foliage to the younger, actively growing leaves. The older leaves (the ones lower on the stem of the tree) become yellow first, while the new leaves remain green.
- *Phosphorus*: The entire seedling is stunted, especially during early growth. Depending on the species, the leaves may become dull green, yellow or purpletinged. The purpling of leaves is a classic symptom, but sometimes there are no colour differences in leaves, so visual diagnosis is not always reliable. The purple colour should not be confused with new leaves that often appear purple or red when they first flush out.
- *Potassium*: Symptoms appear in older leaves first. These start to yellow at the edges, and have some green at the base. Later, leaf edges turn brown and may crinkle or curl and small necrotic (dead) spots may appear. Plants may wilt, even though sufficient water is available in the substrate. When deficiencies are severe, leaves will die. Calcium: This is difficult to detect because signs include slow growth, and dieback of bud or root tips. Seedlings will have stubby little roots with brownish discoloration. The problem is most common in very acidic soils. A well-developed root system with many fine root hairs is important for calcium uptake.
- *Magnesium*: This nutrient is commonly deficient in coarse-structured

soils and in acidic soils. Uptake may be blocked if there is too much potassium in the soil. Like nitrogen, magnesium is a mobile nutrient, so deficiency symptoms show up in the older leaves first. These leaves show a very characteristic yellowing between the veins or ribs, and they appear streaked.

- *Sulphur*: Plants will be slightly stunted. This is not a mobile nutrient, so the symptoms show up on younger leaves which are initially light green, but eventually develop scorched and curled margins. Dry areas can form along the margins and then spread inward to the leaf midrib.

## FERTILIZER MANAGEMENT PRACTICES IN PLANT

### RATES

Nitrogen fertilizer rates are determined by the crop to be grown, yield goal, and quantity of nitrogen that might be provided by the soil. Rates needed to achieve different yields with different crops vary by region, and such decisions are usually based on local recommendations and experience. The quantity of nitrogen supplied by the soil is determined by the quantity of nitrogen released from the soil organic matter, that released by decomposition of residues of the previous crop, any nitrogen supplied by previous applications of organic waste, and any nitrogen carried over from previous fertilizer applications.

Such contributions can be determined by taking nitrogen credits (expressed in lb/acre) for these variables. For example, corn following alfalfa usually requires less additional nitrogen than corn following corn, and less nitrogen fertilizer is needed to reach a given yield goal when manure is applied. As with rates, credits are usually based on local conditions.

Soil testing is being suggested more often as an alternative to taking nitrogen credits. Testing soils for nitrogen has been a useful practice in the drier regions of the Great Plains for many years, and in that region fertilizer rates are often adjusted to account for $NO_3^-$ found in the soil prior to planting. Over the past 10 years, there has been some interest in testing corn fields for $NO_3^-$ in the more humid regions of the eastern US and Canada, utilizing samples taken in late spring, after crop emergence, rather than before planting. This strategy, the pre-side dress nitrogen soil test (PSNT), has received a great deal of publicity and seems to provide some indication of whether additional side dressed nitrogen is needed or not. The test is still in the experimental stage in most states, and interpretations based on local research and experience are essential.

### FERTILIZER PLACEMENT

Placement decisions should maximize availability of nitrogen to crops and minimize potential losses. A plant's roots usually will not grow across the root zone of another plant, so nitrogen must be placed where all plants have direct access to it. Broadcast applications accomplish this objective. Banding does also, when all crop rows are directly next to a band. For corn, banding anhydrous ammonia or UAN in alternate row middles is usually as effective as banding in each middle, because all rows have access to the fertilizer.

Moist soil conditions are necessary for nutrient uptake. Placement below the soil surface can increase nitrogen availability under dry conditions because roots are more likely to find nitrogen in moist soil with such placement. Injecting side dressed UAN may produce higher corn yields than surface application in years when dry weather follows side dressing. In years when rainfall occurs shortly after application, subsurface placement is not as critical. Subsurface placement is normally used to control nitrogen losses.

Anhydrous ammonia must be placed and sealed below the surface to eliminate direct volatilization losses of the gaseous ammonia. Volatilization from urea and UAN solutions can be controlled by incorporation or injection. Incorporating of urea materials (mechanically or by rainfall shortly after application) is especially important in no-till and turfgrass situations where volatilization is aggravated by large amounts of organic material on the soil surface.

Applying small amounts of "starter" nitrogen as UAN in herbicide sprays, however, is usually of little concern. Placing nitrogen with phosphorus often increases phosphorus uptake, particularly when nitrogen is in the $NH_4^+$ form and the crop is growing in an alkaline soil. The reasons for the effect are not completely clear, but may be due to nitrogen increasing root activity and potential for phosphorus uptake, and nitrification of $NH_4^+$ providing acidity which enhances phosphorus solubility.

**TIMING OF NITROGEN APPLICATIONS**

Timing has a major effect on the efficiency of nitrogen management systems. Nitrogen should be applied to avoid periods of significant loss and to provide adequate nitrogen when the crop needs it most. Wheat takes up most of its nitrogen in the spring and early summer, and corn absorbs most nitrogen in mid-summer, so ample availability at these times is critical.

If losses are expected to be minimal, or can be effectively controlled, applications before or immediately after planting are effective for both crops. If significant losses, particularly those due to denitrification or leaching, are anticipated, split applications, where much of the nitrogen is applied after crop emergence, can be effective in reducing losses. Fall applications for corn can be used on well-drained soils, particularly if the nitrogen is applied as anhydrous ammonia amended with N-Serve; however, fall applications should be avoided on poorly drained soils, due to an almost unavoidable potential for significant denitrification losses.

When most of a crop's nitrogen supply will be applied after significant crop growth or positioned away from the seed row (anhydrous ammonia or UAN banded in row middles), applying some nitrogen easily accessible to the seedling at planting ensures that the crop will not become nitrogen deficient before gaining access to the main supply of nitrogen.

## MINIMIZING NITROGEN FERTILIZER LOSSES

The major mechanisms for nitrogen fertilizer loss are denitrification, leaching, and volatilization. Denitrification and leaching occur under very wet soil conditions, while volatilization is most common when soils are only moist and are drying.

## ANHYDROUS AMMONIA

Ammonia is a basic industrial material and is the material from which other nitrogen fertilizers are made. It has been manufactured on a commercial scale since 1913, and the process used to produce it, originally developed by Fritz Haber, has changed little since that time.

In the Haber process, gaseous $H_2$ and $N_2$ are combined under high temperature and pressure conditions to produce ammonia. Ammonia production is an energy intensive process, not only because of its temperature requirements, but also because it requires large quantities of purified $H_2$ that is usually derived from natural gas, or methane ($CH_4$).

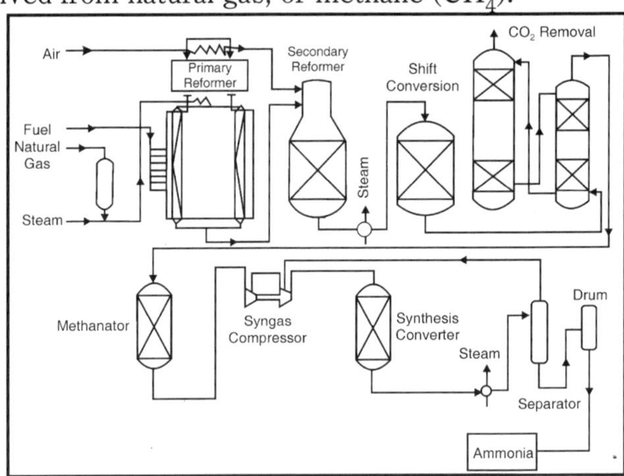


**Fig.** Production of Ammonia

Naphtha and coal are also used as feed stocks in some parts of the world. Hydrogen is produced from methane ($CH_4$) in a multi-step process. First, a "synthesis gas" is produced by "reforming" methane and steam to hydrogen and carbon monoxide (CO) over a catalyst at 1300-1600° F. This synthesis gas is mixed with air in a second reforming reaction over another catalyst, producing a mixture of $N_2$, $H_2$, CO, $CO_2$ and steam.

The carbon-monoxide and steam then undergo a final "shift conversion", producing carbon dioxide and more hydrogen. This final synthesis gas is then purified to remove $CO_2$ that would poison the ammonia-synthesis catalyst. Any remaining traces of carbon oxides are removed by absorption into other materials or are converted back to methane over a methanation catalyst.

The purified $N_2$-$H_2$ synthesis gas is compressed to pressures of 2000-5000 psig (depending on process) and passed over an iron catalyst where ammonia is formed. The resulting gas mixture is cooled to condense and remove ammonia. Unreacted products are recycled through the synthesis converter. The resulting liquid ammonia is then stored in insulated tanks.

## AMMONIUM NITRATE

Nitrate Ammonia can be oxidized in air to produce nitric acid ($HNO_3$). This nitric acid can then be neutralized with more ammonia to produce a solution that is typically 83 percentammonium nitrate and 17 per cent water.

This solution can be used to produce nitrogen fertilizer solutions or can be processed further to produce solid ammonium nitrate. The ammonium nitrate solution is concentrated to 96-99 per cent $NH_4NO_3$ and the "melt" transported to the top of a prilling tower, where it is sprayed into a rising air stream.

The droplets crystallize and condense into hard, spherical "prills" that are dried, cooled, and sized for shipment. Ammonium nitrate can also be granulated by several processes.

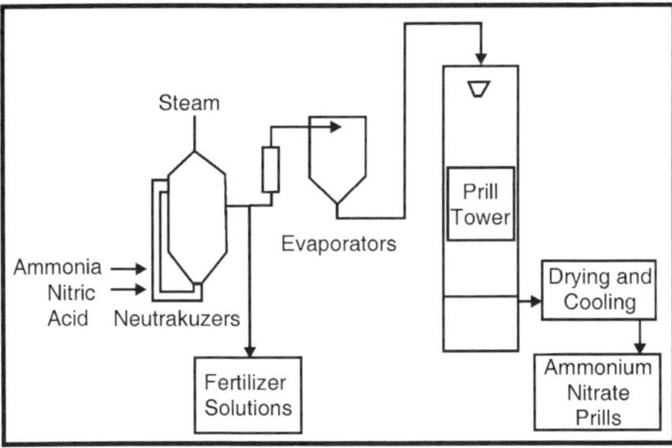


**Fig.** Production of Ammonium

## UREA

Pressurized carbon dioxide gas and heated liquid ammonia are mixed in a condenser to form a mixture containing an intermediate compound, ammonium carbamate. The mixture flows to a reactor where it is dehydrated to convert a portion of the carbamate to urea. This mixture then passes through a stripper, where a stream of carbon dioxide removes unreacted ammonia and recycles it back to the condenser. In the stripper, unreacted ammonium carbamate is decomposed back to ammonia and carbon dioxide, and is also recycled to the condenser. The remaining urea solution is removed from the stripper and purified. The solution can then be used to make nitrogen fertilizer solutions, or can be concentrated and used to produce solid urea in prills or granules.

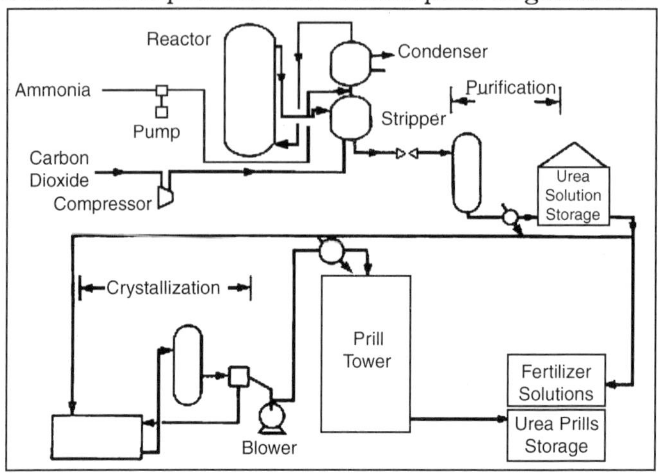


**Fig.** Production of Urea

## UAN SOLUTIONS

Urea-ammonium-nitrate solutions are made by mixing the final solutions produce during production of urea and ammonium nitrate. Mixing in various ratios allows production of materials with different nitrogen concentrations, from 19 to 32 per cent.

## AMMONIUM SULFATE

Ammonium sulfate is a byproduct of the steel and caprolactam industries, made by neutralizing waste sulfuric acid with ammonia. Solid ammonium sulfate is removed from the solution by heating the solution to boiling, evaporating water and concentrating the salt which eventually crystallizes out. The crystal slurry is dewatered in a centrifuge and the wet crystals dried in warm air.

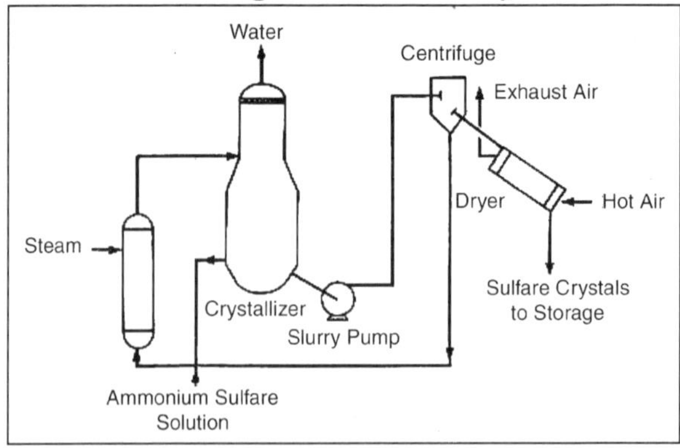


**Fig.** Ammonium Sulfate Crystallization

## AMMONIUM PHOSPHATES

These materials are produced by neutralizing wet process phosphoric acid ($H_3PO_4$) with anhydrous ammonia to produce ammonium orthophosphates (diammonium phosphate (DAP), mono-ammonium phosphate (MAP)) or by neutralizing concentrated "superphosphoric" acid with ammonia to produce ammonium polyphosphates. The orthophosphates are generally granulated as dry materials, while the polyphosphates are used as fluid materials.

## WATER QUALITY CONCERNS INVOLVING NITROGEN

Nitrogen fertilizer use is currently linked to some water quality problems involving surface and underground drinking water supplies. Current public health regulations in the United States and other countries require water suppliers to notify customers when $NO_3^-$ nitrogen concentrations in their water exceed 10 micrograms per litre.

This is because ingestion of large amounts of $NO_3^-$ by very young infants can lead to development of a potentially fatal, but treatable, condition called methemoglobinemia. Nitrate ingestion is usually not harmful to older children or adults. Nearly all natural water contains some $NO_3^-$ at some time during the year and in some regions water has always shown elevated $NO_3^-$ concentrations. However, concentrations may rise temporarily in streams draining watersheds in intensive agricultural production in some years, or may become elevated in groundwater in agricultural regions. Pollution of individual wells can often be traced to improper well construction, coupled with a concentrated loading, such as feedlot run-off.

Elevated concentrations in streams or aquifers are often due to excessive nitrogen applications (often as manure) or failure of the crop to efficiently utilize the nitrogen applied. High$NO_3^-$ concentrations in water, logically, often occur in years following droughts. While droughts, and their effects, cannot be predicted in advance, following best management practices for nitrogen use will help ensure that most $NO_3^-$ concerns do result from abnormal weather rather than improper management decisions.

## COMMERCIAL NITROGEN FERTILIZERS OF PLANT

At one time the fertilizer industry relied on organic or direct-mined materials as sources of nitrogen. Today, virtually all nitrogen materials are manufactured, usually from ammonia. Such materials are less expensive, more concentrated, and just as plant available as the organics used in the past.

## READILY SOLUBLE MATERIALS

### ANHYDROUS AMMONIA ($NH_3$) AND AQUA AMMONIA

Anhydrous ammonia is 82 per cent nitrogen and is a gas at normal atmospheric temperatures and pressure. It is stored as a liquid under pressure and is injected several inches below the soil surface, where it vaporizes and dissolves in the soil water to form $NH_4^+$ ions.

Ammonia (not ammonium) is harmful or toxic to living organisms, and applications near living plants may cause temporary injury. Applications too close to seeds or seedlings may cause stand problems. Though some soil sterilization occurs in the immediate zone of ammonia application, the effect is temporary, and soil life is restored in the zone within a few weeks of application.

Aqua ammonia, a solution of ammonia in water (usually 21 per cent nitrogen), is sometimes used as a fertilizer. It is also injected underground, due to the ammonia's tendency to volatilize from the solution.

## AMMONIUM NITRATE ($NH_4NO_3$)

Ammonium nitrate is usually used as a solid material with an analysis of up to 34 per cent nitrogen. It contains both $NH_4^+$ and $NO_3$-forms of nitrogen, and is used as a source of nitrogen in many blends of liquid and dry fertilizers, as well as being applied directly. Pure ammonium nitrate is very hygroscopic and can be explosive under certain conditions; however, present fertilizer grades of the material are specially conditioned, and when stored and handled properly, pose no problem or hazard.

## UREA [$CO(NH_2)_2$]$_2$

Urea is a manufactured, organic compound containing 46 per cent nitrogen that is widely used in solid and liquid fertilizers. It has relatively desirable handling and storage characteristics, making it the most important solid nitrogen-fertilizer material, worldwide. It may contain small concentrations of a toxic decomposition product, biuret; however, urea manufactured using good quality control practices rarely contains enough to be of agronomic significance. Urea is converted to ammonium carbonate by an enzyme called urease when applied to soil.

Ammonium carbonate is an unstable molecule that can break down into ammonia and carbon dioxide. If the ammonia is not trapped by soil water, it can escape to the atmosphere. This ammonia volatilization can cause significant losses of nitrogen from urea when the fertilizer is applied to the surface of warm, moist soils, particularly those covered with plant residues (no-till) or those drying rapidly. Relatively high surface pH also aggravates nitrogen volatilization from urea.

## NON-PRESSURE NITROGEN SOLUTIONS

These materials are generally composed of urea and/or ammonium nitrate dissolved in water. The most popular are urea-ammonium-nitrate (UAN) solutions. Nitrogen concentrations generally range from 19-32 per cent. These are very versatile materials and are often used as carriers for herbicides or applied in irrigation water. They can be injected, or banded on the soil surface, to minimize potential volatilization losses of the urea component in no-till or surface application situations. Spraying over live foliage can cause severe tissue burn, but the problem can often be minimized by applying in narrow streams rather than by broadcasting. Because the solubility of the nitrogen salts in these solutions decreases as temperature declines, these materials are subject to "salting out". This limits storage and use of more concentrated formulations in cooler seasons and regions.

## AMMONIUM SULFATE [$(NH_4)_2SO_4$]

Ammonium sulfate is a by-product of many industrial processes and contains 20 per cent nitrogen and 24 per cent sulfur. It is not a widely used material, due to its relatively low nitrogen content, but may become more popular as sulfur deficiencies become more wide spread. It is not as prone to nitrogen volatilization as urea, but is slightly more sothan ammonium nitrate, particularly on alkaline soils.

## AMMONIUM PHOSPHATES

Ammonium phosphates are manufactured in a wide variety of forms and formulations including mono-ammonium phosphate (MAP), diammonium phosphate (DAP) and several ammonium polyphosphates. These materials are used mainly as carriers for phosphorus, but can also be significant in nitrogen management programmes, particularly when they are used as starter and row-applied fertilizers.

When used in a starter programme, DAP (at any rate) is usually placed several inches from the seed to avoid damage due to any ammonia released as the material dissolves.

## NITRATE SALTS

Several salts of nitrate, including sodium nitrate ($NaNO_3$, 16 per cent nitrogen), calcium nitrate ($Ca(NO_3)_2$, 15.5 per cent nitrogen), and potassium nitrate ($KNO_3$, 14 per cent nitrogen) are available, but are not widely used due to relatively high cost and low analysis. They are used mainly in specialty situations, such as production of horticultural crops.

## SLOW-RELEASE MATERIALS

There are a number of materials available that release nitrogen slowly into the soil rather than very shortly after application as do those listed above. They are useful in situations where the producer does not want high levels of soluble nitrogen in the system at any one time, due to potentials for excessive losses or excessively rapid plant growth. Such situations occur in container-based horticultural production and inturfgrass management, among others.

Inorganic compounds, such as magnesium ammonium phosphate, and organic compounds, such as urea-formaldehyde (urea form) and isobutylidene diurea (IBDU),are often used to supply nitrogen at slow but fairly predictable rates. Nitrogen release from these compounds is slowed by very low solubility in water, and release from the organic materials is usually related to rate of microbial degradation. Simpler nitrogen compounds, such as urea, can be coated with less soluble materials, including polymers and waxes, and nitrogen release controlled by the rate at which the coating deteriorates.

Sulfur-coated urea (SCU), made by spraying urea granules with molten sulfur, is a common example of such a product, though more are becoming available. All of these products are relatively expensive sources of nitrogen, and their use is restricted to specialty markets and use on high-value crops.

## NITROGEN IN PLANTS

Plants are surrounded by the nitrogen (N) in our atmosphere. Every acre of the earth's surface is covered by thousands of pounds of this essential nutrient, but because atmospheric gaseous nitrogen is present as almost inert nitrogen ($N_2$) molecules, this nitrogen is not directly available to the plants that need it to grow, develop and reproduce. Despite nitrogen being one of the most abundant elements on earth, nitrogen deficiency is probably the most common nutritional problem affecting plants world wide.

Healthy plants often contain 3-4 per cent nitrogen in their above ground tissues. These are much Higher concentrations than those of any other nutrient except carbon, hydrogen and oxygen, nutrients not of soil fertility management concern in most situations. Nitrogen is an important component of many important structural, genetic and metabolic compounds in plant cells. It is a major component of chlorophyll, the compound by which plants use sunlight energy to produce sugars from water and carbon dioxide (*i.e.* photosynthesis).

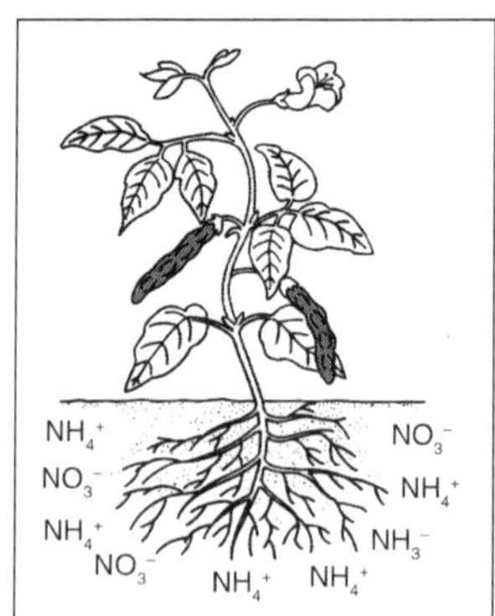


It is also a major component of amino acids, the building blocks of proteins. Some proteins act as structural units in plant cells while others act as enzymes, making possible many of the biochemical reactions on which life is based.

Nitrogen is a component of energy-transfer compounds, such as ATP (adenosine triphosphate) which allow cells to conserve and use the energy released in metabolism. Finally, nitrogen is a significant component of nucleic acids such as DNA, the genetic material that allows cells (and eventually whole plants) to grow and reproduce. Nitrogen plays the same roles (with the exception of photosynthesis) in animals, too. Without nitrogen, there would be no life as we know it.

## SOIL NITROGEN

Soil nitrogen exists in three general forms - organic nitrogen compounds, ammonium ($NH_4^+$) ions, and nitrate ($NO_3^-$) ions. At any given time, 95-99 per cent of the potentially available nitrogen in the soil is in organic forms, either in plant and animal residues, in the relatively stable soil organic matter or in living soil organisms, mainly microbes such as bacteria.

This nitrogen is not directly available to plants, but some can be converted to available forms by microorganisms. Avery small amount of organic nitrogen may exist in soluble organic compounds, such as urea, that may be slightly available to plants.

The majority of plant-available nitrogen is in the inorganic (sometimes called mineral nitrogen) $NH_4^+$ and $NO_3^-$ forms. Ammonium ions bind to the soil's negatively-charged cation exchange complex (CEC) and behave much like other cations in the soil.

Nitrate ions do not bind to the soil solids because they carry negative charges, but exist dissolved in the soil water, or precipitated as soluble salts under dry conditions. Some $NH_4^+$ and $NO_3^-$ may also exist in the crystal structure of certain soil minerals, and maybe quite available; however, such nitrogen is important in only a few soils.

## NATURAL SOURCES OF SOIL NITROGEN

The nitrogen in soil that might eventually be used by plants has two sources–nitrogen containing minerals and the vast storehouse of nitrogen in the atmosphere. The nitrogen in soil minerals is released as the mineral decomposes. This process is generally quite slow and contributes only slightly to nitrogen nutrition on most soils.

On soils containing large quantities of $NH_4^-$ rich clays (either naturally occurring or developed by fixation of $NH_4$ added as fertilizer), however, nitrogen supplied by the mineral fraction may be significant in some years. Atmospheric nitrogen is thought to be a major source of nitrogen in soils. In the atmosphere it exists in the very inert $N_2$ form and must be converted before it becomes useful in the soil. This conversion is accomplished two ways. Some $N_2$ is oxidized to $NO_3$ by lightning during thunderstorms. The $NO_3$ dissolves in raindrops and falls into the soil. The quantity of nitrogen added to the soil in this manner is directly related to thunderstorm activity, but most areas probably receive no more than 20 lb nitrogen/acre per year from this source.

Some microorganisms can utilize atmospheric $N_2$ to manufacture nitrogenous compounds for use in their own cells. This process, called biological nitrogen fixation, requires a great deal of energy; therefore, free-living organisms that perform the reaction, such as *Azotobacter*, generally fix little nitrogen each year (usually less than 20 lb nitrogen/acre), because food energy is usually scarce. Most of this fixed nitrogen is released for use by other organisms upon death of the microorganism. Bacteria such as *Rhizobia*, that infect (nodulate) the roots of, and receive much food energy from, legume plants can fix much more nitrogen per year (some well over 100 lb nitrogen/acre). When the quantity of nitrogen fixed by *Rhizobia* exceeds that needed by the microbes themselves, it is released for use by the host legume plant. This is why well-nodulated legumes do not often respond to additions of nitrogen fertilizer. They are already receiving enough from the bacteria.

## NITROGEN TRANSFORMATIONS AND LOSSES IN SOILS

Nitrogen can go through many transformations in the soil. These transformations are often grouped into a system called the "nitrogen cycle", which can be presented in varying degrees of complexity. Because microorganisms are responsible for most of these processes, they occur very slowly, if at all, when soil temperatures are below 50 °F, but their rates increase rapidly as soils become warmer. The heart of the nitrogen cycle is the conversion of inorganic to organic nitrogen, and vice-versa. As microorganisms grow, they remove $NH_4^+$ and $NO_3^-$ from the soil's inorganic, available-nitrogen pool, converting it to organic nitrogen in a process called immobilization.

When these organisms die and are decomposed by others, excess $NH^{4+}$ can be released back to the inorganic pool in a process called mineralization. Nitrogen can also be mineralized when microorganisms decompose a material containing more nitrogen than they can use at one time, materials such as legume residues or manures. Immobilization and mineralization are conducted by most microorganisms, and are most rapid when soils are warm and moist, but not saturated with water.

The quantity of inorganic nitrogen available for crop use often depends on the amount of mineralization occurring, and the balance between mineralization and immobilization. Ammonium ions ($NH_4^+$) not immobilized or taken up quickly by higher plants are usually converted rapidly to $NO_3^-$ ions by a process called nitrification. This is a two step process, during which bacteria called Nitrosomonas convert $NH_4^+$ to nitrite ($NO_2^-$), and then other bacteria, Nitrobacter, convert the $NO_2^-$ to $NO_3^-$. This process requires a well aerated soil, and occurs rapidly enough that one usually finds mostly $NO_3^-$ rather than $NH_4^+$ in soils during the growing season. The nitrogen cycle contains several routes by which plant-available nitrogen can be lost from the soil. Nitrate-nitrogen is usually more subject to loss than is ammonium nitrogen. Significant loss mechanisms include leaching, denitrification, volatilization, and crop removal. The nitrate form of nitrogen is so soluble that it leaches easily when excess water percolates through the soil. This can be a major loss mechanism in coarse-textured soils where water percolates freely, but is less of a problem in finer-textured, more impermeable soils, where percolation is very slow.

These latter soils tend to become saturated easily, and when microorganisms exhaust the free oxygen supply in the wet soil, some obtain it by decomposing $NO_3^-$. In this process, called denitrification, $NO_3^-$ isconverted to gaseous oxides of nitrogen or to $N_2$ gas, both unavailable to plants. Denitrification can cause major losses of nitrogen when soils are warm and remain saturated for more than a few days.

Losses of $NH_4^+$ nitrogen are less common and occur mainly by volatilization. Ammonium ions are basically anhydrous ammonia ($NH_3$) molecules with an extra hydrogen ($H^+$) attached. When this extra $H^+$ is removed from the $NH_4$ ion by another ion such as hydroxyl ($OH^-$), the resulting $NH_3$ molecule can evaporate, or volatilize from the soil.

This mechanism is most important in high pH soils that contain large quantities of $OH^-$ ions. Crop removal represents a loss because nitrogen in the harvested portions of the crop plant are removed from the field completely. The nitrogen in crop residues is recycled back into the system and is better thought of as immobilized rather than removed. Much is eventually mineralized and may be reutilized by a crop.

## NITROGEN NEEDS OF AND UPTAKE BY PLANTS

Plants absorb nitrogen from the soil as both $NH_4^+$ and $NO_3^-$ ions, but because nitrification is so pervasive in agricultural soils, most of the nitrogen is taken up as nitrate. Nitrate moves freely towards plant roots as they absorb water. Once inside the plant $NO_3^-$ is reduced to an $NH_2$ form and is assimilated to produce more complex compounds. Because plants require very large quantities of nitrogen, an extensive root system is essential to allowing unrestricted uptake. Plants with roots restricted by compaction may show signs of nitrogen deficiency even when adequate nitrogen is present in the soil.

Most plants take nitrogen from the soil continuously throughout their lives and nitrogen demand usually increases as plant size increases. A plant supplied with adequate nitrogen grows rapidly and produces large amounts of succulent, green foliage. Providing adequate nitrogen allows an annual crop, such as corn, to grow to full maturity, rather than delaying it. A nitrogen-deficient plant is generally small and develops slowly because it lacks the nitrogen necessary to manufacture adequate structural and genetic materials.

It is usually pale green or yellowish, because it lacks adequate chlorophyll. Older leaves often become necrotic and die as the plant moves nitrogen from less important older tissues to more important younger ones. On the other hand, some plants may grow so rapidly when supplied with excessive nitrogen that they develop protoplasm faster than they can build sufficient supporting material in cell walls. Such plants are often rather weak and may be prone to mechanical injury. Development of weak straw and lodging of small grains is an example of such an effect.

## ORGANIC NITROGEN IN SOILS

The supply of available nitrogen in soils is often supplemented by nitrogen released from soil organic matter or organic materials added to soils (manure, residues of forage legumes, etc.). The quantity of nitrogen released from any of these materials depends on the composition of the material, particularly its ratio of carbon to nitrogen, and the weather.

When microorganisms decompose materials such as manure and alfalfa residue, that have C/N ratios of 10:1 to 20:1 (lb carbon per lb nitrogen), net mineralization occurs, because the microbes are releasing nitrogen from the residue faster than they can utilize it. When they are decomposing residues such as cornstalks or wheat straw, that have much larger C/N ratios, they must take some extra mineral nitrogen from the soil to utilize all of the carbon they are digesting, so net immobilization occurs.

The rate at which these processes occur depends on weather and soil conditions. When soils are warm and moist, decomposition proceeds rapidly, and nitrogen released from legume residues or manures may be significant, but when soils are cold and wet, or very dry, nitrogen release may be very much less than expected.

There are situations in which nitrogen applications can be reduced or eliminated when corn follows good stands of alfalfa or another legume. However, the nitrogen relationships involved here are so site-specific and weather-related that local recommendations should be consulted before making decisions regarding nitrogen credits. Legume species, stand density, age, soil drainage, and timing of operations are all factors to be considered.

The timing and rate of manure application should also be considered when calculating nitrogen credits, and again, due to influences of weather and soil, local guidelines should be used.

Manure can be applied at rates that supply all nitrogen, phosphorus, and potassium to a crop, but such rates usually cause significant over-applications of phosphorus and potassium. Over-applications of phosphorus are important water quality concerns in many regions, so manure applications should be at rates low enough to avoid this over-application. At such low application rates, supplementary nitrogen-fertilizer applications are usually needed in crops such as corn.

## CROP RESPONSE TO MICRONUTRIENTS

**Table. Relative Responsiveness of Selected Crops to Micronutrients**

| Crop | B | Cu | Mn | Zn |
|---|---|---|---|---|
| Alfalfa | High | Med | Low | Med |
| Apples | High | Med | Low | Med |
| Sugar Beet | High | Low | Low | Med |
| Cabbage | Med | Low | Med | Low |
| Citrus | Med | High | Med | Med |
| Clover | Med | Med | Low | Med |
| Corn | Med | Low | Low | High |
| Cotton | High | Low | High | Med |
| Grain Sorghum | Low | Med | Med | High |
| Grass | Low | Low | Low | Med |
| Lettuce | Med | High | High | Med |
| Oat | Low | Med | High | Med |
| Onion | Low | High | High | High |
| Peach | Med | Med | Med | High |
| Peanut | High | Low | Med | Low |
| Pecan | Med | Low | Low | High |
| Potato, Irish | Low | Med | Med | High |
| Potato, Sweet | High | Low | High | Med |
| Rye | Low | Low | Low | Med |
| Soybean | Low | Med | High | Med |
| Tobacco | Med | Low | Med | Med |
| Tomato | High | High | Med | Med |
| Wheat | Low | High | High | Low |

## MICRONUTRIENT FERTILIZER SOURCES

**Table. Micronutrient Fertilizer Sources**

| Source Element | Solubility in $H_2O$ | % |
|---|---|---|
| | **Boron** | |
| $H_3BO_3$ | Soluble | 17 |
| $Na_2B_4O_7.5H_2O$ | Soluble | 20 |
| $Na_2B_4O_7.10H_2O$ | Soluble | 11 |
| $Ca_2B_6O_{11}.5H_2O$ | Slightly soluble | 10 |
| | **Copper** | |
| $CuSO_4.5H_2O$ | Soluble | 25 |
| CuO | Insoluble | 50-75 |
| | **Iron** | |
| $FeSO_4.7H_2O$ | Soluble | 20 |
| FeHEDTA | Soluble | 5-9 |
| FeEDDHA | Soluble | 6 |

| | | |
|---|---|---|
| | **Manganese** | |
| $MnSO_4.4H_2O$ | Soluble | 24 |
| MnO | Insoluble | 41-68 |
| Mn oxysulfate | Variable | 30-50 |
| | **Molybdenum** | |
| $Na_2MoO_4.2H_2O$ | Soluble | 39 |
| $(NH_4)_2MoO_4$ | Soluble | 49 |
| $MoO_3$ | Soluble | 66 |
| | **Zinc** | |
| $ZnSO_4.H_2O$ | Soluble | 36 |
| $ZnSO_4$ - $NH_3$ complex | Soluble | 10-15 |
| ZnO | Insoluble | 60-78 |
| Zn oxysulfate | Variable | 18-50 |
| ZnEDTA | Soluble | 6-14 |

Micronutrient sources vary considerably in their physical state, chemical reactivity, cost, and availability to plants. Some of the commonly used micronutrient sources are listed in table.

*The four main classes of micronutrient sources are*:

1. Inorganic products
2. Synthetic chelates
3. Natural organic complexes
4. Fritted glass products (frits).

## INORGANIC SOURCES

Inorganic sources include oxides and carbonates, and metallic salts such as sulfates, chlorides, and nitrates. The sulfates are the most common of the metallic salts and are sold in crystalline or granular form. An ammoniated $ZnSO_4$ solution also is used in polyphosphate starter fertilizers. Oxides of manganese and zinc also are commonly used, and are sold as fine powders and in granular form.Because oxides such as ZnO and MnO are water insoluble, their immediate effectiveness for crops is rather low in granular form.

Also, the available divalent form of manganese in MnO will oxidize to the unavailable tetravalent form of manganese, so there is very little residual availability of manganese fertilizers for succeeding crops. Thus, agronomic effectiveness of granular MnO may be rather low.

Since manganese in $MnO_2$ already is in the unavailable form, it should not be used as a manganese fertilizer. Oxysulfates are oxides, usually industrial by-products, which have been partially acidulated with sulfuric acid, and generally are sold in granular form.

The percentage of water-soluble manganese or zinc in oxysulfates is directly related to the degree of acidulation by sulfuric acid. Research results have shown that about 35 to 50 per cent of the total zinc in granular zinc-oxysulfate should be in water-soluble form to be immediately effective for crops.

Similar results would be expected for manganese-oxysulfate. Inorganic sources usually are the least costly sources per unit of micronutrient, but they may not always be the most effective for crops.

## SYNTHETIC CHELATES

These sources are formed by combining a chelating agent with a metal through coordinate bonding. Stability of the metal-chelate bond affects availability to plants of the micronutrient metals—copper, iron, manganese, and zinc. An effective chelate is one in which the rate of substitution of the chelated micronutrient for other cations in the soil is quite low, thus maintaining the applied micronutrient in chelated form.

Relative effectiveness for crops per unit of micronutrient as soil-applied chelates may be from two to five times greater than that of inorganic sources, while chelates costs per unit of micronutrient may be five to 100 times higher. Several types of chelates are sold, so relative effectiveness values depend on the sources of chelates and inorganic products being compared.

## NATURAL ORGANIC COMPLEXES

These complexes are made by reacting metallic salts with some organic by-products of the wood pulp industry or related industries. Several types of these complexes are the lignosulfonates, polyflavonoids and phenols. The types of chemical bonding of the metals to the organic components are not well understood. Some bonds may be coordinate as in the chelates, but other types of chemical bonds also may be present. While natural organic complexes are less costly per unit of micronutrient, they usually are less effective than synthetic chelates. They also are more readily decomposed by microorganisms in soil. These sources are more suitable for foliar sprays and mixing with fluid fertilizers.

## FRITS

Fritted glassy products (frits) in which solubility is controlled by particle size and changes in matrix composition. Micronutrient concentrations vary from 2 to 25 per cent, and more than one micronutrient may be included in a fritted product.

Fritted micronutrients generally are used only on sandy soils in regions of high rainfall were leaching occurs. This class of materials is more appropriate for maintenance programmes than for correcting severe micronutrient deficiencies. Therefore, frits only have a small share of the micronutrient market.

## APPLICATION WITH MIXED FERTILIZERS

The most common method of micronutrient application for crops is soil application. Recommended application rates usually are less than 10 lb/acre (on an elemental basis), so uniform application of micronutrient sources separately in the field is difficult. Therefore, both granular and fluid NPK fertilizers are commonly used as carriers of micronutrients. Including micronutrients with mixed fertilizers is a convenient method of application and allows more uniform distribution with conventional application equipment. Costs also are reduced by eliminating a separate application.

*Four methods of applying micronutrients with mixed fertilizers are*:

1. Incorporation during manufacture
2. Bulk blending with granular fertilizers
3. Coating onto granular fertilizers
4. Mixing with fluid fertilizers

## INCORPORATION WITH GRANULAR FERTILIZERS

Incorporation during manufacture results in uniform distribution of micronutrients throughout granular NPK fertilizers. Because the micronutrient source is in contact with the mixed fertilizer components under conditions of high temperature and moisture, the rate of chemical reactions which may reduce the plant availability of some micronutrients is increased.

For example, acid decomposition of ZnEDTA or any synthetic chelate may occur if they are mixed with phosphoric acid before ammoniation during manufacture, which results in reduced plant availability of the micronutrient. Immediate plant availability of applied zinc in granular ammoniated phosphates also decreases with the level of water-soluble zinc in these products.

## BULK BLENDING WITH GRANULAR FERTILIZERS

Bulk blending of micronutrients with granular NPK fertilizers is a common practice in the U. S. The main advantage is that fertilizer grades can be produced which will provide the recommended micronutrient rates for a given field at the usual fertilizer application rates. The main disadvantage is that segregation of nutrients can occur during the blending operation and with subsequent handling.

Segregation results in no uniform application, which is critical with micronutrients since their application rates are quite low. Segregation can be minimized by properly matching particle sizes of micronutrient sources with those of the NPK components of the blend. Mechanical devices to minimize coning and segregation of the materials during handling and storage are available. Blending of various sized fertilizer particles results in no uniform application because of segregation in the applicator during transport and spreading operations.

## COATING GRANULAR FERTILIZERS

Coating powdered micronutrients onto granular NPK fertilizers decreases the possibility of segregation, which is the main disadvantage of bulk blending micronutrients with mixed fertilizers. Fertilizer solutions are preferred as binding agents because the fertilizer grade is not decreased so much as with use of water, oils and waxes. Some binding materials are unsatisfactory because they do not maintain the micronutrient coatings during bagging, storage, and handling. This results in segregation of the micronutrient sources from the granular NPK components. Agronomic effectiveness of micronutrients coated onto soluble granular NPK fertilizers should be similar to that with incorporation during manufacture. This method of micronutrient application is not commonly used because of the extra costs associated with coating.

## FLUID FERTILIZERS

Mixing micronutrients with fluid fertilizers has become a popular method of application, especially in the U. S. Clear liquids are commonly used as starter fertilizers for row crops and some micronutrients, especially zinc sources, are easily applied with these fluids. Solubility of some micronutrient sources is higher in polyphosphate fertilizers such as 10-34-0 than in orthophosphate clear liquids. Micronutrients also may be applied with nitrogen solutions such as UAN, but solubility of many sources is rather low. Compatibility tests should be made before tank mixing operations of micronutrients with fluid fertilizers are attempted; otherwise, problems could occur when incompatible sources are mixed. Suspension fertilizers also are used as micronutrient carriers. Oxides also can be applied with suspensions since complete solution is not required.

## FOLIAR SPRAYS

Foliar sprays are widely used to apply micronutrients, especially iron and manganese, for many crops. Soluble inorganic salts generally are as effective as synthetic chelates in foliar sprays, so the inorganic salts usually are chosen because of lower costs. Suspected micronutrient deficiencies may be diagnosed with foliar spray trials with one or more micronutrients.

Correction of deficiency symptoms usually occurs within the first several days and then the entire field could be sprayed with the appropriate micronutrient source. Inclusion of sticker-spreader agents in the spray is suggested to improve adherence of the micronutrient source to the foliage. Caution should be used because of leaf burn due to high salt concentrations or inclusion of certain compounds in foliar sprays.

*Advantages of foliar sprays are*:

- Application rates are much lower than for soil application;
- A uniform application is easily obtained;
- Response to the applied nutrient is almost immediate so deficiencies can be corrected during the growing season.

Low residue foliar sprays of manganese and zinc have been used to correct deficiencies of citrus and other fruit crops, but sprays which will discolor the fruit should be avoided.

*Disadvantages of foliar sprays are*:

- Leaf burn may result if salt concentrations of the spray are too high;
- Nutrient demand often is high when the plants are small and leaf surface is insufficient for foliar absorption;
- Maximum yields may not be possible if spraying is delayed until deficiency symptoms appear;
- There is little residual effect from foliar sprays. Application costs will be higher if more than one spray is needed, unless they can be combined with pesticide spray applications.

## MICRONUTRIENT RATES AND EXAMPLE RESULTSS

### BORON

Recommended application rates of boron are rather low (0.5 to 2 lb/acre), but should be carefully followed because the range between boron deficiency and toxicity in most plants is narrow. Uniform application of boron in the field is very important for the above reason. Boronated NPK fertilizers (those containing boron sources incorporated at the factory) will insure a more uniform application than most bulk blended fertilizers. Foliar sprays also insure a rather uniform application, but costs generally are higher.

In an experiment in Arkansas, cotton yields were increased by 490 and 584 lb/acre at boron rates of 0.3 and 0.5 lb/acre, respectively. Response of cotton in Tennessee to boron application is shown below. Without applied boron, cotton yields decreased with increasing soil pH. Yields were increased at all soil pH levels when boron was applied at a rate of 0.5 lb/acre. Soil tests should be included in boron fertilization programmes, first to assess the level of available boron and later to determine possible residual effects (buildup). The most common soil test for boron is the hot-watersoluble test. This test is more difficult to conduct than most other micronutrient soil tests, but most boron response data have been correlated with it.

### COPPER

Recommended copper rates range from 3 to 10 lb/acre as $CuSO_4$ or finely ground CuO. Residual effects of applied copper are very marked, with responses being noted up to eight years after application.

Because of these residual effects, soil tests are essential to monitor possible copper accumulations to toxic levels in soils where copper fertilizers are being applied. Plant analyses also can be used to monitor copper levels in plant tissues. Copper applications should be decreased or discontinued when available levels increase beyond the deficiency range.

### IRON

Soil applications of most iron sources generally are not effective for crops, so foliar sprays are the recommended application method. Spray applications of a 3 to

4 per cent $FeSO_4$ solution at 20 to 40 gallons/acre are used to correct iron deficiencies.

The application rate should be high enough to wet the foliage. More than one foliar application may be required for correction of iron chlorosis. Inclusion of a sticker-spreader agent in the spray is suggested to improve adherence of the spray to the plant foliage for increased iron absorption by the plant.

## MANGANESE

Recommended application rates range from 2 to 20 lb/acre of manganese, generally as $MnSO_4$. Application rates of MnO would be similar if applied as a fine powder or in NPK fertilizers.

Band application of manganese sources with acid-forming fertilizers results in a more efficient use of applied manganese because the rate of oxidation of applied manganese to the unavailable tetravalent form (as in $MnO_2$) is decreased. There are no residual effects of applied manganese for the same reason, so annual applications are needed.

Foliar spray applications of $MnSO_4$ also are used and require lower rates than soil applications.

**Table. Soybean Response to Manganese**

| lb/acre Mn | Yield, bu/acre |
|---|---|
| 0 | 13 |
| 2.5 | 24 |
| 5.0 | 26 |
| 10.0 | 33 |

Results given below show that soybean yields were more than doubled when 10 lb of manganese/acre was applied as $MnSO_4$ to a poorly drained sandy soil in Georgia. Cotton also responded to applied Manganese in Georgia, as shown below.

**Table. Cotton Response to Manganese**

| lb/acre Mn | Yield, lb/acre | Increase, lb/acre |
|---|---|---|
| 0 | 2,657 | — — |
| 2.5 | 2,857 | 200 |

## MOLYBDENUM

Recommended molybdenum rates are much lower than those for the other micronutrients, and uniform application is very important. Broadcast application of molybdenized phosphate fertilizers prior to planting or to pastures has been used to correct molybdenum deficiencies. Soluble molybdenum sources also can be sprayed on the soil surface before tillage to obtain a uniform application.

**Table. Soybean Response to Molybdenum**

| | Mo Applied, oz/acre | |
|---|---|---|
| Soil pH | 0 | 1 |
| | —Yield, bu/acre—- | |
| 5.3 | 25.7 | 45.0 |
| 5.7 | 34.3 | 45.0 |
| 6.3 | 41.8 | 46.1 |

Seed treatment is the most common method of molybdenum application. Molybdenum sources are coated onto the seed with a sticking agent and/or conditioner. This method insures a uniform application and sufficient amounts of molybdenum can be seed coated to provide sufficient molybdenum. Data in the following table show the effectiveness for soybean in Georgia of seed-coated molybdenum at a rate of one ounce of molybdenum/acre. Soybean yields without applied molybdenum increased with increases in soil pH, but not as high as those with seed-applied molybdenum at each soil pH level.

## ZINC

Recommended rates of zinc generally range from 1 to 10 lb/acre. Band or broadcast applications are used, but foliar applications also are effective. Band applications of zinc sources with starter fertilizers is a common practice for row crops. Foliar sprays of a 0.5 per cent $ZnSO_4$ solution applied at a rate of 20 to 30 gallons/acre also will supply sufficient zinc, but several applications may be necessary.

As with copper, residual effects of applied zinc are substantial, with responses found at least 5 years after application. Because of these residual effects, soil test levels of available zinc generally increase after several applications. Many states have reduced their recommended zinc application rates because of these residual effects. Crop response to several zinc sources each banded with a 10-34-0 starter fertilizer at zinc rates up to 3 lb/acre for corn in Nebraska is shown below. Results show that ZnEDTA was much more effective at the lower zinc rates, but all zinc sources were about equally effective at the highest zinc rate.

Micronutrients are as important as the primary and secondary nutrients in plant nutrition. However, the amounts of micronutrients required for optimum nutrition are much lower. Micronutrient deficiencies are widespread because of increased nutrient demands from the more intensive cropping practices. Soil tests and plant analyses are excellent diagnostic tools to monitor the micronutrient status of soils and crops. Visual deficiency symptoms of these nutrients also are well recognized in most economic crops. Micronutrient recommendations are based on soil and plant tissue analyses, the type of crop and expected yield, management level, and research results. Numerous micro-nutrient fertilizers are on the market.

These sources are classified as inorganic, synthetic chelates, natural organic complexes, and fritted glasses. Some industrial by-products are used as micronutrient fertilizers because of their lower cost. Micronutrient sources vary considerably in physical form, chemical reactivity, cost, and relative availability to plants. While most micronutrient fertilizers are applied to soils, foliar sprays also are used on tree crops and some vegetables. Because recommended rates usually are low, most micronutrients are applied with NPK fertilizers by incorporation during manufacture or bulk blending with granular fertilizers, or by mixing with fluid fertilizers just before application to soil. Choice of micronutrient source depends on the method of application, compatibility with the NPK fertilizer, convenience of application, and the relative agronomic effectiveness and cost per unit of micronutrient.

## SOIL COMPOSITION IN PLANT NUTRIENTS

Soil is made up of three main things: clay, humus and sand. There are also many small organisms that live in the soil, and most of these are useful to the plants. *Clay* Clay is made from the breakdown and recombination of silicate rocks. It is made up of alternating layers of silicon oxides then aluminium oxides, with various cations such as $Ca^{2+}$ loosely bound in between the layers.

Anions adsorb onto the oxide surfaces, with doubly and triply charged cations sticking better than singly charged ones. Clays are the main source of nutrients in the soil. *Humus* This is any organic matter in the soil - *i.e.* the products of the decay of plants and animals. It is mostly made up of aromatic compounds. Over time it breaks down to carbon dioxide and water so it needs to be continually replaced. Humus is important in regulating the amount of water in the soil. *Sand* This is solid particles of ground up rock. A small amount of sand is necessary to ensure the correct water content in the soil.

## SOURCES OF PLANT NUTRIENTS

The nutrients available in a given soil ultimately depends on the rock from which the soil was made. If the plants grown from this soil die and decay where they have grown then their nutrients are recycled. However, if the plants have been grown for agriculture then they are removed from the area in which they have grown and their nutrients cannot be recycled. So in soils that are used for cropping, essential nutrients continually have to be replaced.

## PHOSPHORUS

The phosphorus originally comes from the *weathering* of the parent material, which releases phosphate ions into the soil solution. This phosphate is *adsorbed* onto soil constituents, forming "reservoirs" of various types and capacities, some of which release phosphate ions into the soil solution easily, while others strongly bind phosphate. Plants absorb phosphate from the soil solution, and incorporate it into their tissues as they grow. When the plants die, decay processes slowly convert the complex phosphorus compounds in the plant tissues back to simple phosphate ions (*mineralization*) and the cycle repeats.

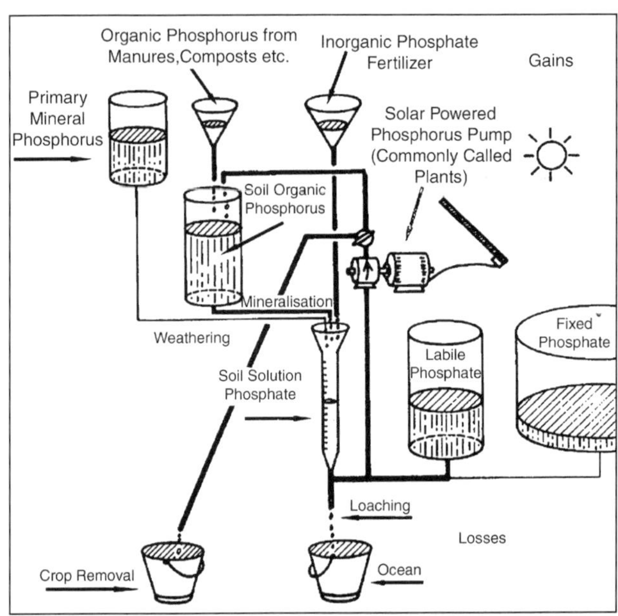


Under "natural" conditions the losses by leaching are to some extent made up by weathering of more parent minerals to release fresh phosphate. If a crop is removed from the area then the soil will contain less phosphate, the phosphate reservoirs will gradually become depleted and the soil will become less fertile.

The phosphate content of the soil can be increased by addition of phosphorus as "inorganic" phosphate which goes straight into the soil solution (if soluble), or dissolves slowly (if sparingly soluble). Alternatively, phosphorus can be added as complex organic phosphorus compounds, such as are present in plant or animal materials, which release their phosphorus slowly as they decay. Phosphorus is bound so firmly to soil that leaching losses into waterways are generally small, though erosion which carries phosphorus-rich soil particles into the waterways can encourage the growth of algae and weed in rivers and lakes (*eutrophication*).

## NITROGEN

Nitrogen is one essential nutrient which behaves quite differently. There is virtually no nitrogen in rocks; most of the world's nitrogen is in the air, which is 80 per cent dinitrogen, $N_2$. However, most plants are unable to make use of dinitrogen, and can only absorb nitrogen from the soil solution in the form of nitrate or ammonium ions. The dinitrogen of the air must be converted into nitrate or ammonia by the process of *nitrogen fixation* before it becomes available to plants. This nitrogen fixation can occur by either a physical process, or a biological process. The physical process occurs whenever air is heated to a high temperature. The dinitrogen and dioxygen of the air combine to form nitric oxide

$$N_2(g) + O_2(g) \rightarrow 2NO(g).$$

This reaction occurs naturally in lightning strikes and forest fires. Human activity causes many more high temperature processes, such as in internal combustion and jet engines, thermal power plants, steel and cement works, etc, and has greatly increased the amount of nitrogen fixation.

The nitric oxide formed undergoes further reactions in the atmosphere, and is finally converted into nitric acid, which is washed out by rain and enters the soil as nitrate ions. Too much of this type of nitrogen fixation can cause atmospheric smog and acid rain.

Biological processes in the soil reduce the nitrate to ammonium ions. Biological fixation occurs in certain micro-organisms which have the ability to convert dinitrogen into ammonia, using the enzyme *nitrogenase*, for which the element molybdenum is required. Some of these micro-organisms live freely in the soil (*e.g.* the bacterium *azotobacter*). These bacteria fix the nitrogen for their own needs, but when they die and decay the nitrogen in their cells becomes available for plants. Other bacteria form an alliance (*symbiosis*) with certain plants (*e.g rhizobia* with legumes).

The plant provides a safe place for the bacteria to live (in nodules on its roots), and a food supply (carbohydrate obtained from photosynthesis). In return the bacteria provide the plant with fixed nitrogen as ammonium ions. Much of New Zealand agriculture is dependent on this relationship. Clover in the pasture has bacteria in root nodules which fix nitrogen, so the clover plant is rich in nitrogen. This is eaten by herbivores, and the excess nitrogen is excreted onto the soil in the animal's urine, when it becomes available to the grasses which form the bulk of the pasture. Farmers would have to use much more nitrogen fertilizer to achieve the same amount of pasture growth without the assistance of the nitrogen fixing bacteria in nodules on clover roots.

The life of fixed nitrogen in the soil is short. Both nitrate and ammonium ions are bound weakly by ion-exchange processes, so they are both leached out by rainwater, and finish up in ground water in wells and in rivers. Nitrate and ammonium ions encourage the growth of weed and algae in rivers and lakes, and when these are used for water supplies the toxicity of nitrate ions can cause health problem if the concentrations become too high. Nitrogen is also lost from the soil by the activities of *denitrifying bacteria*.

These convert the nitrate and ammonium ions back to dinitrogen, or more usually, to nitrous oxide ($N_2O$). This is a *greenhouse gas* and so increasing the amount of fixed nitrogen in the soil indirectly increases the amount of nitrous oxide in the atmosphere and potentially contributes to the raising of the temperature of the earth. Farming practices which lead to high levels of nitrogen in the soil can therefore contribute to environmental problems.

## OTHER NUTRIENTS

*Cycles* similar to that described for phosphorus can be drawn for other major nutrients such as potassium, calcium or magnesium, and for trace nutrients, such as cobalt, molybdenum or selenium, which are derived from the soil parent material.

## REPLACEMENT OF PLANT NUTRIENTS

This can be done either using manure or compost (which are relatively poor sources of nutrients) or a synthetic fertilizer. Synthetic fertilizers are usually used to replace nitrogen (in the form of urea, ammonium sulfate, ammonium nitrate or diammonium hydrogen phosphate), phosphorous, sulfur and potassium. The nitrogen is obtained from the air, the phosphorous is mined as $Ca_3(PO_4)_2$, the sulfur is mined in elemental form and reacted with oxygen to form sulfate, and the potasium is obtained from seawater. Other minor nutrients are also sometimes replaced using specialised artificial fertilizers.

## COMPOSTS AND MANURES

In *organic* agriculture nutrients are replenished using only natural materials, though what is and what is not natural is often debated. The main "organic" nutrient sources are composts and manures.

## COMPOSTS

Composts are formed from plant residues, and have much the same chemical composition as the plants from which they are made. Composts are *not* a rich source of plant nutrients, and need to be used in large amounts if they are the only source of nutrient replenishment.

Their main use is to boost the amount of soil organic matter (humus), and composts should be regarded primarily as *soil conditioner*s. There is also a philosophical problem that if compost is made from "imported" plant material, then the area from which the material was obtained has been depleted in nutrients.

As example, continual use of lawn mowings to make compost depletes the lawn area of nutrients, which need to be replenished to keep the lawn healthy. Therefore, while all available plant material should be converted into compost to maintain humus levels and to recycle nutrients, this does not really solve the nutrient deficit problem. An additional problem arises if a region is deficient in some particular nutrient. The plants (and animals) grown in the region will also be deficient in this nutrient, and so also will be composts made from plants grown in the area.

## MANURES

Manures are usually the faeces of herbivores. In the digestion process the animal absorbs as much as it can of the nutrient elements for its own needs, so the residue finally excreted consists of plant material depleted in nutrients.

The nutrients not incorporated into the animal are excreted in the animal's urine, which is a better source of plant nutrients than the faeces. As with compost, manures are useful mainly as soil conditioners. They are not a rich source of plant nutrients, and need to be used in large amounts if they are the only source of nutrient replenishment.

The old gardening advice "to dig in six inches of well rotted stable manure" is fine in principle, but not very practicable for most modern gardeners, and there remains the problem that the area where the animal was feeding has been depleted of the nutrients. Animal residues, such as "blood and bone" form another type of "natural" fertilizer. These can contain high levels of nutrients, which are generally released slowly into the soil solution as the complex compounds are broken down, which has advantages. Again these nutrients have been "mined" from the soil where the animals fed.

## FERTILIZERS

The only answer to the "nutrient mining" problem is to bring fresh nutrients into the area from some richer source, and fortunately natural processes have concentrated the major nutrients in locations around the world. All "synthetic" nitrogen fertilizers obtain their nitrogen from the atmosphere. This is "fixed" using the *Haber process*, in which dihydrogen and dinitrogen at high temperature and pressure are combined over a catalyst, which speeds up the reaction forming ammonia:

$$N_2(g) + 3H_2(g) \rightarrow 2NH_3(g).$$

The hydrogen required for the Haber process is normally made using methane separated from natural gas. Methane is reacted with water and air at high temperatures over catalysts to form a mixture of dihydrogen, dinitrogen (from the air) and carbon dioxide. Carbon dioxide and water are removed (the carbon dioxide is used later to make urea), leaving dihydrogen and dinitrogen in the correct proportion to combine to form ammonia (unbalanced equation)

$$CH_4(g) + H_2O(g) + [O_2(g) + N_2(g)] \rightarrow [N_2(g) + 3H_2(g)] + CO_2(g).$$

The ammonia obtained from the Haber process is converted into a form more useful as fertilizer. It can be combined with carbon dioxide to form urea, which is chemically identical to the urea excreted in the urine of animal:

$$2NH_3(g) + CO_2(g) \rightarrow O{=}C(NH_2)_2(s) + H_2O(l)$$

or combined with sulfuric, nitric or phosphoric acid to form ammonium sulfate, ammonium nitratc or diammonium hydrogen phosphate $(NII_4)_2(IIPO_4)$. The nitric acid is formed by oxidizing ammonia over a catalyst

$$NH_3(g) + 2O_2(g) \rightarrow HNO_3(g) + H_2O(l)$$

## PHOSPHORUS

Phosphorus is a relatively abundant element in the earth's crust, and occurs naturally in marine deposits of *phosphate rock*, as calcium phosphate, $Ca_3(PO_4)_2$, or the minerals *apatite*, $CaF_2.3Ca_3(PO_4)_2$, or *hydroxy apatite*, $Ca(OH)_2.3Ca_3(PO_4)_2$ (hydroxy apatite is the mineral that forms bones and teeth). The phosphate rock from the Pacific region is found on coral islands where flocks of sea birds roost.

It is thought that the phosphate in their excreta, derived from their fishy diet, combines over time with coral to form the phosphate rock. Some phosphate rock is sufficiently soluble so that the phosphorus from the finely ground rock is available to plants. Other phosphate rock is too insoluble for direct use, and is commonly treated with sulfuric acid to make a mixture of calcium sulfate and the more soluble calcium hydrogen phosphate, called *superphosphate*.

$$Ca(OH)_2.3Ca_3(PO_4)_2 + H_2SO_4 \rightarrow CaSO_4 + Ca(HPO_4)_2$$

Phosphate rock can also be reacted with excess sulfuric acid to make phosphoric acid, $H_3PO_4$, which can be converted into ammonium hydrogen phosphate, $(NH_4)_2(HPO_4)$ for use in soluble fertilizers, or into sparingly soluble magnesium ammonium phosphate $Mg(NH_4)(PO_4)$ (magamp) for use as a slow-release fertilizer.

## SULFUR

Sulfur is also an essential nutrient, and some areas of New Zealand are sulfate deficient. Use of superphosphate, or ammonium, potassium or calcium sulfate (gypsum) as fertilizer supplies the needed sulfate. The sulfuric acid needed to make superphosphate is obtained by burning sulfur to form sulfur dioxide

$$S(s) + O_2(g) \rightarrow SO_2(g)$$

which is oxidised over a catalyst to form sulfur trioxide,

$$2SO_2(g) + O_2(g) \rightarrow SO_3(g)$$

which reacts with water to form sulfuric acid,

$$SO_3(g) + H_2O(l) \rightarrow H_2SO_4(aq).$$

The sulfur is obtained from deposits of elementary sulfur found in volcanic areas of the world, or is extracted from natural gas, which in some places contains hydrogen sulphide. Sulfur dioxide is released into the atmosphere from volcanoes. This is ultimately converted into sulfuric acid in the atmosphere; this is washed out by rain, so providing a "natural" source of sulphur.

The 1995 eruption of Ruapehu "topdressed" much of the central North Island with economically useful amounts of sulphate. Sulfur dioxide is also released into the atmosphere when coal or oil containing sulfur is burned in power stations, and the resulting sulfuric acid in the atmosphere causes "acid rain" in some industrialized areas.

## POTASSIUM

Potassium is the other main nutrient needed, and this is obtained from sea water, directly or indirectly. When sea water evaporates sodium chloride (common salt) crystallises first. If the liquid remaining (bittens) is evaporated potassium chloride crystallises. In some places this occurred over geological times when oceans were isolated by geological processes and then evaporated, leaving large deposits of salt (which crystallised first) and of potassium chloride (which crystallised later).

The potassium chloride can be used directly as a fertilizer, and is satisfactory for use at low rates, but this has the disadvantage that chloride is also added, and many plants are not very tolerant of high levels of chloride. It is therefore usual to convert the potassium chloride to potassium sulfate (which also occurs in natural deposits) for use as a fertilizer.

-The so-called "un-natural" inorganic fertilizers are thus derived from the very natural sources of the air (plus natural gas as a hydrogen source) for nitrogen, coral rock plus sea-bird droppings (plus sulfur from volcanoes or "sour" natural gas to make superphosphate) for phosphorus, and the ocean for potassium.

## SOIL ANALYSIS

If soil has an inappropriate composition for a given plant species then that plant will not thrive (or maybe will not grow at all) in that soil. To prevent this being a problem, samples of soil are often analysed so that nutrients can be added before seeds are planted. Soil is commonly analysed for potassium, magnesium, calcium, nitrogen and phosphorous (the exact choice of elements to monitor depends on which plants are to be grown) and also for pH.

Some of these tests are so simple they can be carried out at home and others require specialised equipment. Plants grow in the thin upper layer of the Earth's crust known as soil. Soil is formed over very long times from igneous or sedimentary rock, volcanic ash, sand or peat. With the exception of peat, these are all silicate minerals, so soil is largely made up of silicates and organic material known as humus. Plants use this soil for anchorage and as a source of nutrients and water.

## WHY PLANTS NEED SOIL

Green plants are very effective photochemical factories: they use the energy of light to combine carbon dioxide from the air with water from the soil to make carbohydrates:

$$6CO_2(g) + 6H_2O(l) + \text{energy} \rightarrow C_6H_{12}O_6(aq) + 6O_2(g).$$

This reaction occurs in the chloroplasts of all green plants, using the green pigment chlorophyll to trap the energy from sunlight. The basic building blocks of glucose are then used to build up more complex carbohydrates such as cellulose. However, plants need more than just carbon dioxide, water and sunlight to keep alive.

The elemental composition of a 'typical' plant. It can be seen that the composition of plants is not much different from that of animals, and includes significant amounts of nitrogen and phosphorous, as well as trace amounts of many other elements. Although there is enormous variation between the abundance of the most common element, hydrogen, and the least common, molybdenum, both are very important.

For example, plants can only use nitrogen in the form of the ammonium ion. Nitrates are converted to ammonium ions by the enzyme *nitrate reductase*, and as this enzyme cannot function without molybdenum, the presence of molybdenum is essential if the plant is to receive nitrogen in a form that it can use. These trace elements, generally called plant nutrients, normally come from the soil in which the plant is growing.

The plants absorb their nutrients from the soil solution (the film of water which covers all the particles in moist soil), usually in very simple forms. A typical composition of the main constituents of the soil solution from which plants must obtain all their requirements for growth. It is possible to grow plants without soil, using *hydroponics*. The plant is grown in a solution which contains all the required plant nutrients, in the correct proportions.

It is possible to determine just how much of any particular nutrient a plant needs by starting with a balanced nutrient solution, and then varying the amount of each nutrient present, one nutrient at a time. When insufficient of the nutrient is present the plant will not thrive.

As the amount of nutrient is increased the growth improves, until all that the plant can utilize is present, when adding more of that nutrient produces no additional growth. If the amount of the nutrient is increased still further, the excess may eventually cause toxic effects, and growth is retarded. For some nutrients, *e.g.* potassium, plants can tolerate a large excess before ill effects occur; for others, *e.g.* boron, the optimum range is narrow.

## THE COMPOSITION OF SOIL

Apart from peat, which is derived from plant material, soil is composed of silicate minerals of various types, composed largely of oxygen and silicon, with smaller amounts of aluminium, iron, sodium, potassium, calcium and magnesium, and variable and much smaller amounts of all the other elements.

As rainwater saturated with carbon dioxide percolates through silicate minerals it dissolves out the simple soluble cations, such as $Na^+$, $K^+$ and $Ca^{2+}$, and anions such as $Cl^-$, in the process of *weathering*. The residue consists largely of oxides of silicon, aluminium and iron.

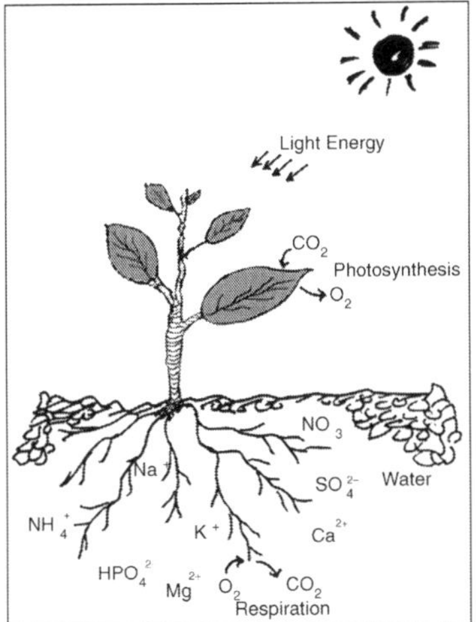


**Fig.** The Plant Environment

**Table. Typical Composition of Soil Solution**

| Cation | Conc./mmol $L^{-1}$ | Anion | Conc./mmol $L^{-1}$ |
|---|---|---|---|
| $Ca^{2+}$ | 10 | $NO_3^-$ | 5 |
| $Mg^{2+}$ | 3 | $SO_4^{2-}$ | 4 |
| $K^+$ | 1 | $Cl^-$ | 2 |
| $Na^+$ | 1 | $HCO_3^-$ | 2 |
| $NH_4^+$ | 0.5 | $HPO_4^{2-}$, $H_2PO_4$ | 0.01 |

Over hundreds to thousands of years this recrystallises as *clay minerals*. These are *aluminosilicate* compounds, composed largely of silicon, aluminium and oxygen, which have *layer structures* built up like stacks of sandwiches.

Silicon atoms, each surrounded by a tetrahedron of oxygen atoms, are linked into sheets by sharing oxygen atoms. Aluminium atoms, each surrounded by an octahedron of oxygen atoms (since aluminium is larger than silicon), similarly form sheets. These sheets combine to form layers by sharing oxygen atoms. In the *kaolin* type of clay minerals one silicon-oxygen sheet is combined with one aluminium-oxygen sheet to form each layer.

These layers are then loosely stacked together like stacks of bread-and butter (—bread-butter—bread-butter—-). In another type (*smectite* type) a central silicon-oxygen sheet has an aluminium-oxygen sheet on each side; these are then loosely stacked like a bread and butter sandwich (—bread-butter-bread—bread-butter-bread—-). The crystals of the clay minerals are very small, typically less than 4 ìm in size (such a particle would contain about 5,000 layers), and so have large total surface areas.The attractive forces between the layers are weak, and for some clay minerals water can diffuse between the layers, making the effective surface area even larger.

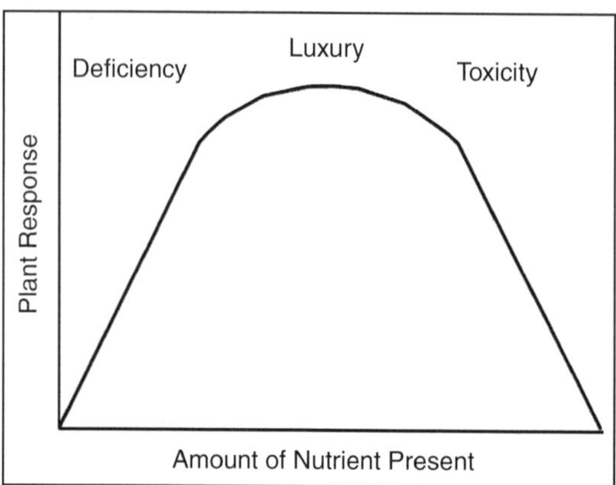


**Fig.** Response of Plant to Amount of One Nutrient

The clay structures are generally not perfect, and "wrong" atoms are frequently incorporated into the layers. Such mistakes can upset the charge balance in the layers, and lead to unbalanced negative charges appearing in the oxygen sheets.

These charges are balanced either by incorporating a hydrogen ion (converting an -O- to -O-H), or by having a cation, such as $Ca^{2+}$, somewhere nearby between the layers. The edges of the layers also have excess negative charges, which have to be similarly balanced. Figure represents a typical clay mineral, with loosely bound cations between the layers and at the surface of the particles. The places where hydrogen or cations are present are called *exchangeable cation sites*.

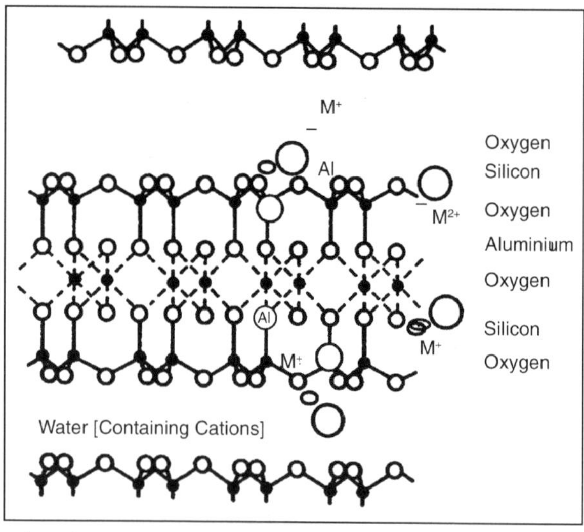


**Fig.** The Structure of a Clay Mineral, Showing Cation Exchangeable Sites

The cations bound by the clay particle can interchange with cations in solution. Clay minerals thus have an ability to loosely bind cations and *exchange* them with other cations in the soil solution. This *cation exchange capacity* of clays is an important property of soils. Figure shows the way in which the cations loosely held in soils can be exchanged for others from the soil solution.

The clay minerals, and also tiny particles of the oxides of iron and aluminium present in soil, have another important property; they adsorb anions on their surfaces. The strength of this anion binding increases with charge, so that singly charged anions, such as $Cl^-$ or $NO_3^-$ are poorly retained by soil, doubly charged $SO_4^{2-}$ is more strongly bound, and triply charged $PO_4^{3-}$ is very strongly bound.

Some New Zealand soils formed from volcanic ash contain a clay mineral called *allophane*. This has a very imperfect aluminosilicate structure in which the layers are thought to curl around to form tiny porous balls about 5 nm in diameter, rather like plastic practice golf balls. These form soils with a very large total surface area—about a rugby-field of surface for every 10 grams of allophane. These soils have the ability to bind phosphate so strongly that it becomes unavailable to plants, and is said to be "fixed". Soils also contain organic matter, called *humus*.

This is the ultimate product of the decay of plant and animal material, and is intimately mixed in with the clay by the actions of the soil's population of organisms. Humus is largely derived from the most resistant of plant material, the lignin from the cell walls.

This contains much aromatic material (six-carbon benzene rings) with attached hydroxy (-OH) and carboxylic acid (-COOH) groups. These can also ionize to form -O- and -COO- groups, and these charges have to be balanced by a cation in the vicinity, so humus also contributes to the ion exchange capacity of the soil. The humus is slowly, continuously, being broken down to simpler compounds, and ultimately to $CO_2$ and $H_2O$ and must continuously be replenished with new plant residues to maintain the soil's "condition".

Without the ability of soils to bind both cations and anions the essential plant nutrients would be rapidly washed out of the soil, and plant life would probably not have developed in the forms we know. The physical properties of a soil are very important in determining its usefulness for gardening or agriculture.

A fertile soil will generally have a balance of clay minerals and organic material, and probably also coarser particles of "sand" as silica or other resistant minerals of the original rock. The humus present improves the physical properties of the soil, and gives it the ability to drain excess water yet retain sufficient moisture for plant growth. The balance of these constituents determines the soil's properties; how well it holds water, and drains, whether it is "sticky" or friable, how well it stands up to repeated cultivation, etc, and so determine its usefulness for agriculture. Soils have a thriving living population, ranging in size from earthworms to single celled organisms, living on plant and animal residues, which are partly responsible for the creation of fertile soil from weathering rock. Most soil organisms are beneficial, though some eat plant roots or cause or transmit plant diseases.

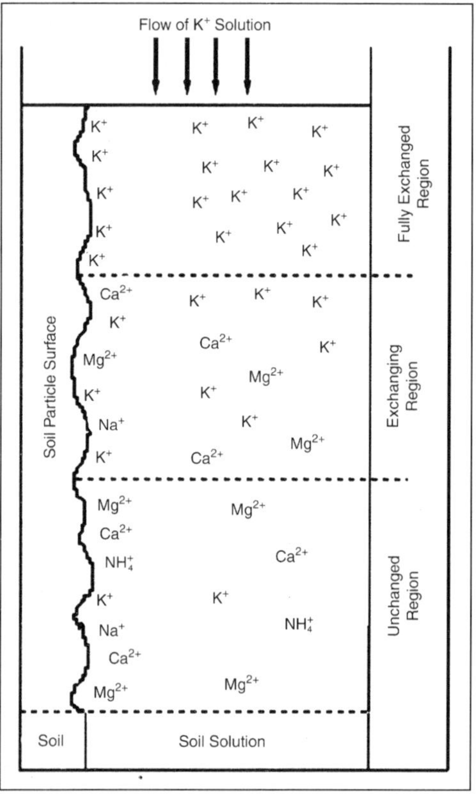


**Fig.** Exchange of Soil-bound Cations by Solution Containing Potassium Ions. The Anions Necessary for Charge Neutralization are Omitted for Clarity

## RETENTION OF CATIONS IN SOILS

Cations are held more strongly (less reversibly) when pH increases from 5 to 7. Cu, Zn, Ni, Cd and other metals become signicantly less soluble and less exchangeable when pH increases from 5 to 7.

*Retention of metals in soil can occur through several processes*:

- Cation exchange (non-specic adsorption),
- Specic adsorption,
- Organic complexation
- Co-precipitation. In a given situation most, if not all, of these processes contribute to metal retention in soils.

In order to maintain electroneutrality, negative charges on solid particles (soil colloids) are balanced by an equal amount of cations; an exchange refers to the exchange between counter-ions balancing the surface negative charge on the soil colloids and ions in soil solution. Such an exchange is reversible, stoichiometric and diffusion controlled.

Moreover, there is a certain degree of selectivity of the adsorbent. The higher the valency of an ion, the greater its replacing power ($H^+$ behaves like a polyvalent cation). By contrast, with greater degrees of hydration, a given ion will exhibit a lower replacing power. Adsorption by cation exchange represents electrostatic binding through the formation of outer-sphere complexes with the surface functional groups. An outer-sphere complex means that at least one molecule of a solvent comes between the functional group and the ion.

Specic adsorption is pH dependent and related to the hydrolysis of the heavy-metal ion. In specic adsorption partly covalent bonds are formed with the lattice ions. Partly covalent bonds are inherently stronger than electrostatic binding involved in the non-specic cation exchange (*e.g.* Zn can be adsorbed on Fe and Al oxides 7 and 26 times more strongly than their corresponding cation exchange capacity at pH 7.6 would imply).

*Metals most able to form hydroxy complexes are specically adsorbed to the greatest extent*:

$$Hg > Pb > Cu >> Zn > Co > Ni > Cd$$

Speciûc adsorption may also include diffusion of metals into mineral interlayer spaces and their xation there. Such diffusion increases with an increase in pH. Organic matter may either increase or decrease availabil-ity of micronutrients, Al and heavy metals.

Reduced availability is due to complexation with humic acids, lignin and other organic compounds of high molecular weight (insoluble precipitates are thus formed). Conversely, increased availability may result from solubilisation and thus mobilisation of metals by low molecular weight organic ligands (*e.g.* short-chain organic acids, amino acids and other organic compounds). Stability constants of chelates with several metals occur with increasing order as:

$$Cu > Fe = Al > Mn = Co > Zn$$

Co-precipitation represents formation of mixed solids by simultaneous precipitation as occurs with Fe and Mn oxides.

## SOIL–PLANT PH

Increasing acidity or alkalinity correspond to logarithmic increase in concentration of $H^+$ and $OH^-$ respectively (vertical bars). Horizontal bars represent relative availability (or toxicity) at any particular pH. Most agricultural soils will be slightly acid (pH around 5.5 to 6.5) and essential nutrients are all readily available within that range. Of particular note, highly acid soils are conducive to both Al and Mn toxicity and to Mo deficiency. Highly alkaline soils are conducive to B toxicity but to Fe, Zn and Mn deficiencies.

The pH value most relevant to soil and plant chemical processes is pH of the soil solution. A soil is acidic if the pH of its aqueous solution phase is <7 and alkaline if that pH exceeds 7. Nutrient element availability varies accordingly, and beyond the range of pH 4–8 plant growth becomes a function of pH *per se*, plus pH effects on nutrient ion availability. In chemical terms, pH represents a measure of $H^+$ activity in a soil solution which is in a dynamic equilibrium with a negatively charged solid phase. $H^+$ ions are strongly attracted to these negative sites and have sufcient power to replace other cations from them. A diffuse layer in the vicinity of a negatively charged surface has higher $H^+$ activity than the bulk soil solution.

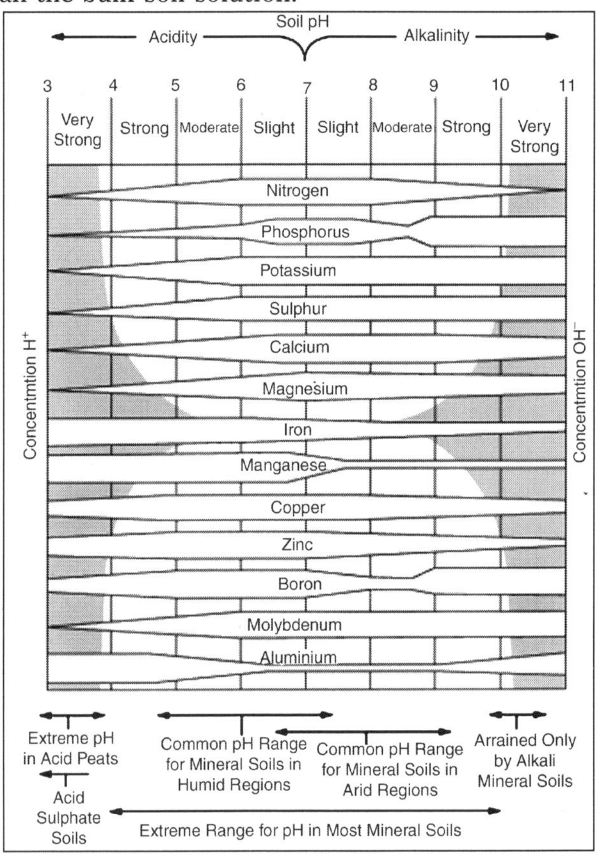


**Fig.** A Highly Diagrammatic Picture of Soil Nutrient Availability (and Element Toxicity) as a Function of pH.

Soil pH varies in time and space. Diurnal fluctuations of as much as one pH unit may occur, as well as spatial variations (horizontal and vertical down the soil prole). Soil pH also varies over seasons. During seasons with low to moderate rainfall when evapotranspiration greatly exceeds precipitation, salts are not being removed by deep percolation and increased salts tend to reduce pH by forcing more of the exchangeable $H^+$ ions into the soil solution. Conversely, during wet seasons, salts are removed from the topsoil and pH goes up. This season to season fluctuation in total salt content should not be confused with long-term soil acidication.

## RELATIONSHIPS BETWEEN PH AND ION TOXICITY

Soil pH is a dominant influence on solubility and therefore availability and potential phytotoxicity of metals. Low pH favours free metal cations and protonated anions, higher pH favours carbonate or hydroxyl complexes. There-fore, availability of micronutrient and toxic ions (which are present in soil solution as cations) increases with increasing soil acidity. By contrast, availability of those present as anions ($MoO_4^{2-}$, $CrO_4^{2-}$, $SeO_4^-$, $SeO_3^-$ and $B(OH)_4^-$) increases with increasing alkalinity.

## RHIZOSPHERE

Plant growth is dependent on availability of water and nutrients in the rhizosphere, the soil–root interface consisting of a soil layer varying in thickness between 0.1 mm and up to a few millimetres depending on the length of root hairs. Availability of nutrients in the rhizosphere is controlled by the combined effects of soil properties and interactions between plant roots and adjacent microorganisms in the surrounding soil.

Chemical conditions in the rhizosphere are usually very much different from those in the bulk soil further away from roots. Root-induced changes in the rhizosphere pH are a result of the balance between $H^+$ and $HCO_3^-$ excretion, evolution of $CO_2$ by respiration and loss of various organic compounds known collectively as root exudates. The balance between $H^+$ and $HCO_3^-$ excretion depends upon the cation/anion uptake ratio.

Greater excretion of $H^+$ accompanies a greater absorption of cations than anions and results in rhizosphere acidication. Conversely, when uptake of anions exceeds uptake of cations, excretion of $HCO_3^-$ exceeds that of $H^+$. The chemical form of soil N (ammonium v. nitrate) is an influential factor for the cation/anion ratio. Ammonium-fed plants take up more cations than anions, and they usually have a more acidic rhizosphere than bulk soil, while nitrate-fed plants take up more anions than cations and show the opposite relationship between rhizosphere and bulk soil pH.

Plant effects on rhizosphere pH also vary with genotype, which can in turn influence nutrient ion availability. Overall, plants and soils must be regarded as interacting components in any ecosystem, and because plants take up more basic than acidic components, any net increase in ecosystem biomass will result in some degree of soil acidication.

## ESSENTIAL MINERAL NUTRIENTS

Solution culture experiments have established a number of facts basic to plant growth on soils—that plant roots must have a supply of oxygen, that solid soil particles and micro-organisms, while sometimes benecial, are not essential, that in addition to C, H and O some 14 or 15 chemical elements are essential for plant growth, and that all of these essential elements may be supplied to plant roots as simple ions of inorganic salts in solution and must be supplied in adequate but non-toxic amounts.

A chemical element is regarded as essential if, in its absence, a plant cannot complete its life cycle. An alternative criterion has also been suggested—that the element is part of an essential plant constituent or molecule. So far, only the rst criterion has been used to establish essentiality although the latter has proved useful as a guide to further research, as, for example, in the association of Mn with laccase activity and Ni with urease activity.

'Essential mineral nutrients' or simply 'essential nutrients' include all those chemical elements which are normally absorbed from the soil solution by higher plants. They exclude C, H, and O which comprise more than 99 per cent of metabolically active leaves and some 90–95 per cent of their dry matter, but, paradoxically, they include N which does not occur in any soil mineral and is supplied to some plants from air.

Essential mineral nutrients comprise less than 1 per cent of fresh mass in active leaves and 5–10 per cent of their dry matter; yet within this small proportion, all 14 or 15 elements must be present in adequate amounts for healthy growth and effective reproduction.

## FUNCTIONS OF NUTRIENTS

A Summary of notional values for concentrations of nutrient elements (on a Dry Mass Basis) considered as just adequate for maximum palnt growth, and accompanying data on the relative number of atoms of those elements witnin a plant (referenced to Mo=l).

A broad distribution between plant requirements for micronutrients (trace elements) and micronutrients is than evident. Mobility form leaves is meant to reflect overall impressions chemical analaysis of nutrient budgets for whole plants and organs.

Historically, essential nutrients have been classied in one of two groups based on amounts required by plants, namely macronutrients or micronutrients, and that convention has been adopted for convenience in Table. This distinction reflects historical sequence and experimental difûculties in discovering essential nutrients. Following advances in nine-teenth century chemistry, essentiality of macronutrients was relatively easy to prove, but essentiality of micronutrients (except Fe) was elusive, requiring great care in eliminating contamination from macronutrient salts, from water and from other environmental sources. The classication remains in common use, and was therefore used in Table but the distinction between macro and micro is somewhat arbitrary. In terms of chemistry and function, an alternative grouping is outlined below.

## GROUP 1: N, S

N and S are covalently bound in organic compounds in reduced states. They are essential components of proteins providing reactive groups for interaction with metabolites and other nutrients. They are generally absorbed as oxyanions which, except for small amounts of sulphate in a few organic com-pounds, must be reduced before use.

## GROUP 2: P, B

P and B are covalently bound in organic compounds in their fully oxidised states. P is present in phospholipids of cell membranes, nucleic acids of chromosomes and a large number of intermediary metabolites including many in which it plays a crucial role in bioenergetics. P is always present as mono- or polyorthophosphate which behaves as a weak acid. B, like P, is thought to function only in its fully oxidised form; B oxyacid is much weaker and is largely undissociated at the pH of cells. B complexes strongly with hydroxyls in adjacent *cis*-diol conguration in sugars and their derivatives such as those in the hemicellulose of cell walls, but key metabolic or structural functions still remain unknown.

## GROUP 3: K, CA, MG, NA, CL

K, Ca, Mg, Na and Cl are present primarily in ionic form as either free ions in solution or as entities reversibly adsorbed by electrostatic forces on charged sites. K, Ca and Mg are required in relatively large amounts while Na and Cl, though often present in large amounts, are required only in trace amounts. Group 3 nutrients function in osmotic adjustment of cell activities and in enzyme activation, probably by modifying the shape and orientation of substrates and enzymes. Ca and to a lesser extent Mg also stabilise cell structures such as membranes and cell walls, probably via their divalent positive charge which forms cross-links with negative charges on cell structures. Some Ca binds strongly and reversibly to the protein calmodulin which in turn activates several enzymes and controls Ca transport. Some Mg is structurally bound in chlorophyll (a strict 1:1 stoichiometry of Mg atoms to chlorophyll molecules is universal)

## GROUP 4: MN, FE, CO, NI, CU, ZN, MO

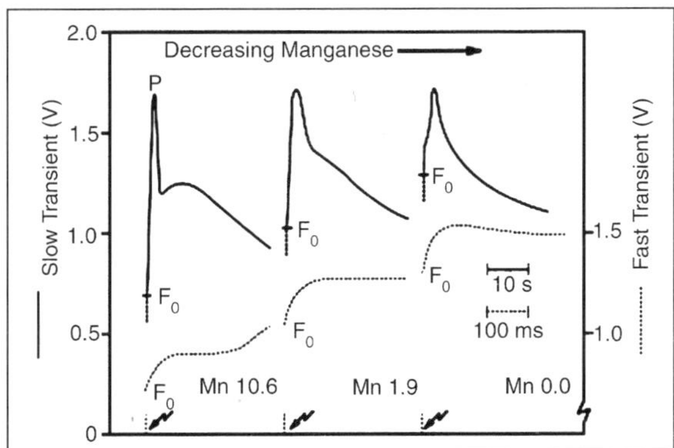


**Fig.** Induction Kinetics for *in Vivo* Chlorophyll *a* Fluorescence Change Dramatically According to Leaf Mn Status of Wheat Leaves.

Constnat yield fluorescence ($F_0$) increases but variable fluorescence ($F_v$; where $F_v = P–F_0$) decreases with decrease in leaf Mn. A ratio of $F_0/F_v$ increases abruptly as leaf Mn drops to a critical level, and provides an early indication of impending Mn deficiency.

Preplanting soil dressings corresponding to transients on leaf samples were, left to right, 10.6, 1.9 and 0.0 kg $MnSO_4$ $ha^{-1}$. Fe, Co, Ni, Cu, Zn, Mo and, to some extent, Mn are tightly bound in proteins as metalloproteins or, in the case of Co, which is only required in plants xing $N_2$, in a coenzyme. All are transition series elements, and with the exception of Zn all show variation in valency state with attendant variation in physiological effect. Group 4 elements govern a wide range of reactions including many oxidation–reductions in which all except Zn can act as electron carriers by undergoing reversible oxidation–reduction.

All are micronutrients, and all except Mn form strong complexes with organic and amino acids. Mn is exceptional in that it resembles Group 3 nutrients in being present largely as a free divalent ion which can replace Mg in the activation of a number of enzymes. But some Mn is tightly bound within the water-splitting apparatus of photo-system II and in superoxide dismutase, an enzyme system that helps dissipate harmful effects of singlet oxygen formation. Mn deciency results in highly charac-teristic induction kinetics for *in vivo* chlorophyll *a* fluorescence and can be used as a diagnostic tool in combination with chemical assay.

## SOIL ACIDITY AND TOXICITIES

Elements essential for plant growth generally occur at soil concentrations that range between decient or adequate for plant growth. Instances do occur, however, where there are excess concentrations of either essential or non-essential elements. These toxicities may occur naturally, as saline soils, in soils developed from parent materials high in heavy metals, or through soil acidication.

Human action can exacerbate natural processes. This has occurred with salinisation of the landscape as a result of clearing native vegetation or through irrigation. Heavy metals have accumulated in the landscape through industrial pollution, through the use of fungicides that contain Cu or the use of superphosphate which may contain Cd. Accelerated acidication has occurred through certain farming practices and is a particularly serious problem on sandy soils with low buffer capacity.

Application of N fertilizers and inclusion of pasture legumes in rotations increase acidication. Also, drainage of some estuarine soils for farming or urban uses, in which S is present in the reduced state (*e.g.* as iron pyrite) under waterlogged conditions (*i.e.* acid sulphate soils), results in oxidation of S and reduction in soil pH.

Certain mine spoils are also acid, particularly those high in iron pyrite, which on oxidation result in acidication of spoil *in situ* or of run-off waters. Acid rain is a further problem, and in industrialised nations, especially in Europe and northeastern USA, there is considerable atmospheric pollution which produces acid rain containing dissolved oxides of N and S. This acid rain may fall long distances away from the source of pollution, and has led to international disputes.

## SOIL ACIDICATION

As soils acidify, plant growth generally declines and there is a decrease in the range of plant species that may be grown. The activity of soil fauna and flora decreases also. The extent of acid soils and the degree to which plant growth is reduced by acid soil factors make these soils of considerable ecological and economic importance.

Soil pH (negative logarithm of the molar activity of hydrogen ions) has been used to dene acid soils. Since pH is measured on a logarithmic scale, it is important to realise that a soil with pH 4 is 10 times more acid than a soil with pH 5! Acid soils may be classied as those soils with a pH of less than 7 (*i.e.* neutral pH).

However, this has little value in practice, since most plant species will only show reduced growth at substantially lower pH. Thus, soils with a pH of less than 5.5 may be regarded as acid (*i.e.* those in which growth of many plant species is adversely affected).

Acid soils occupy about 30 per cent of global land surfaces, and predominate in two major regions of the world: humid temperate forests and humid tropics and subtropics. However, many soils in other areas are acid also. Tropical acid soils alone comprise approximately two billion hectares, or 14 per cent, of the total ice-free area of the world. Approximately 40 per cent of the world's arable soils (*i.e.* those on which crops may be grown) are acid.

Eight regions of Australia have been identied in which agricultural production is reduced by acid soils. In southern Australia, these include the wheat and sheep belt of Western Australia, the southern portion of South Australia, cropping and grazing lands of southeastern South Australia along with the western, central and northeastern districts of Victoria. There are also about one million hectares of highly acidic soils in Tasmania.

Further north in New South Wales, the cropping and grazing regions of the Riverina and southwestern slopes contain many acid soils, as do the tablelands and wet sub-tropical coastal farming regions of New South Wales and Queensland. Soils of the tropical wet coast of Queensland and adjacent intensively cropped and grazed areas are also acid to a considerable extent.

Overall, acid soils are most common where rainfall exceeds 450 mm $year^{-1}$ and more than 80 million hectares of the most productive agricultural land in Australia are acidic, with more than 40 per cent of this land being highly acidic.

Acid sulphate soils of coastal areas may also acidify when drained for crop production or urban use, with consequent effects on estuarine aquatic life. This most noticeably results in sh kills, often following rains after a dry period during which oxidation of soil S has occurred.

Acid soil infertility in Australia is now recognised as an insidious and invisible form of land degradation. Worldwide, soil nutrient decline and acidication are often the most easily visible forms of land degradation. A landmark Australian study showed that soils under subterranean clover pasture acidify, with a decrease from a soil pH of around 6.0 in virgin soils to pH 5.2 where pasture had been grown for more than 30 years.

Subsequent studies have shown that the acid addition rate over extensive areas of New South Wales ranges from near zero to 3–5 kmol $H^+$ $ha^{-1}$ $year^{-1}$, with up to 20 kmol $H^+$ $ha^{-1}$ $year^{-1}$ in some exploitative systems. With an acid addition rate of 4 kmol $H^+$ $ha^{-1}$ $year^{-1}$, there would be a decrease of about 1 pH unit in the surface 30 cm of a sandy loam within 30 years. This would take about 120 years in more strongly buffered clay soils.

## ALUMINIUM AND MANGANESE TOXICITIES

Soil pH has a marked effect on nutrient availability by affecting both solubility of soil minerals present and the ability of plant roots to absorb nutrients. Generally, a decrease in soil pH increases availability of cations, especially essential micronutrients, Fe, Mn, Cu and Zn. In contrast, availability of Mo decreases, often to such an extent that acid soils are decient in plant-available Mo. Availability of P also decreases at low pH via xation with hydrous oxides of Fe and Al, although reactions with Ca can also reduce P availablility around pH 8.5.

## ALUMINIUM TOXICITY

Al toxicity has been recognised for over 60 years. Shortly after the turn of the twentieth century, soil scientists found Al present in solution leached from soils. However, later attention was given to hydrogen clays, and following development of glass electrodes research emphasis was placed on soil pH for many years—an example of technology driving science! Despite publication of research results between the late 1920s and late 1960s, deleterious effects of Al on plant growth did not attract interest again until the early 1970s. Some of those ndings are outlined here.

**Table. Many froms of Al are Potentially Present in Soil Solutions of Acid Soils**

| From | Examples |
|---|---|
| Micro-crystals | Kaolinite, Gibbsite |
| Amorphous Precipitates | $Al_2SiO_5$, $Al(OH)_3$, $AlPO_4$ |
| Inorgaic Polycations | $Al_2(OH)_2^{4+}$, $Al_3(OH)_3^{5+}$ |
| | $AlO_4Al_{12}(OH)_{24}(H_2O)_{24}^{?+}$ |

| | |
|---|---|
| Organic Complexes | Al citrate, Al oxalate<br>Al fulvate, Al humate |
| Inorganic Complexes | $AlF^{2+}$, $AlF_2^{+}$, $AlSO_4^{+}$ |
| Inorganic Monomers | $Al^{3+}$, $AlOH^{2+}$, $Al(OH)_2^{+}$ |

Al is the most common metal in the earth's crust (8 per cent of dry mass) and comprises some 7 per cent of soils, where it has a complex chemistry. Al is an important component of aluminosilicate compounds, including clay particles. As soils acidify, Al dissolves from these solid forms, and enters solution with the potential to then become toxic. Not all Al in the soil solution is toxic, however, providing challenges to biologists to identify those forms that are. Those shown to be toxic to plants include the inorganic Al monomers ($Al^{3+}$, $AlOH^{2+}$, $Al(OH)_2^{+}$) and the inorganic poly-cation$AlO_4Al_{12}(OH)_{24}(H_2O)_{24}^{7+}$ (known as $Al_{13}$). The concentration of Al in the soil solution is often low (<50 $\mu$M), and activities of monomeric Al in the soil solution that reduce root growth have been found to range from 4 to 15 $\mu$M.

A further challenge to analytical chemists is to develop techniques able to discriminate between toxic and non-toxic forms of Al, made all the more difcult because of the low concentration of toxic Al. This concentration has been calculated to be about 0.1 g Al in 1 $m^3$ of soil (*c.* 12 $\mu$M Al or one-third of one part per million in the soil solution).

The same 1 $m^3$ of acid soil would contain a total of 70 kg Al. Solubility of Al decreases dramatically with an increase in soil pH, and the concentration of inorganic monomeric Al in solution is also affected by organic and inorganic anions. Complexes of Al with organic ligands of low relative molecular mass, especially citric and malic acids, are also considerably less toxic. Tolerant plants benet from such chelating mechanisms.

In addition, organic matter increases the concentration of fulvic and humic acids, resulting in the formation of Al fulvate and Al humate. These complexes are considerably less toxic than the inorganic monomeric species.

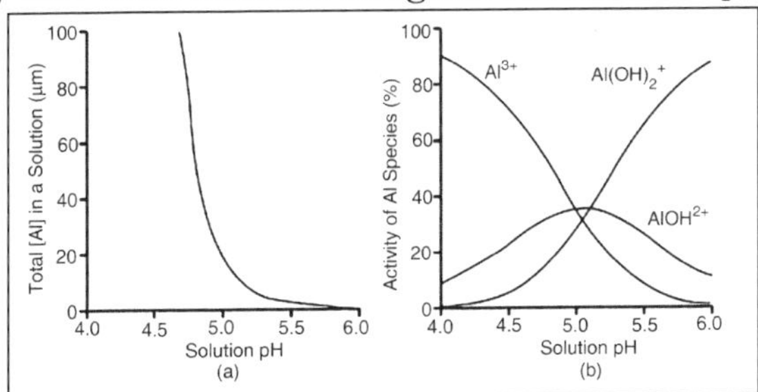


(a) Total concentration of all ionic species of Al in solution as a function of pH. In (b) a solution that initially contains 1000 µM $CaCl_2$ and 100 µM $AlCl_3$ (used to approximate a soil solution) shows a marked decrease in concentration of $Al^{3+}$ from pH 4.0 to pH 6.0, and a change in relative activities of different inorganic monomeric Al species.

Underlying equilibria responsible for such pH effects are summarised below. Values for pK coincide with the pH of a solution where the reaction mixture would be 50 per cent dissociated. Increasing H+ ion concentration due to decreasing pH forces such equilibria towards $Al^{3+}$, thus inducing Al toxicity.

$$Al^{3+} + H_2O \Leftrightarrow AlOH^{2+} + H^+ \quad pK = 5.00$$

$$Al^{3+} + 2H_2O \Leftrightarrow AlOH^{2+} + 2H^+ \quad pK = 10.10$$

Biological effects of ions are often better related to activ-ity in solution rather than to concentration. In addition to soil pH effects on Al solubility, speciation of Al in solution changes with pH. At pH 4.0, most of the inorganic monomeric Al is present as $Al^{3+}$; this decreases with an increase in pH while the activities of the hydroxy Al species ($AlOH^{2+}$, $Al(OH)_2^+$) increase. At alkaline pH, aluminate ions (*e.g.* $Al(OH)_4^-$, $Al(OH)_5^{2-}$) enter into solution but are not toxic to plant roots.

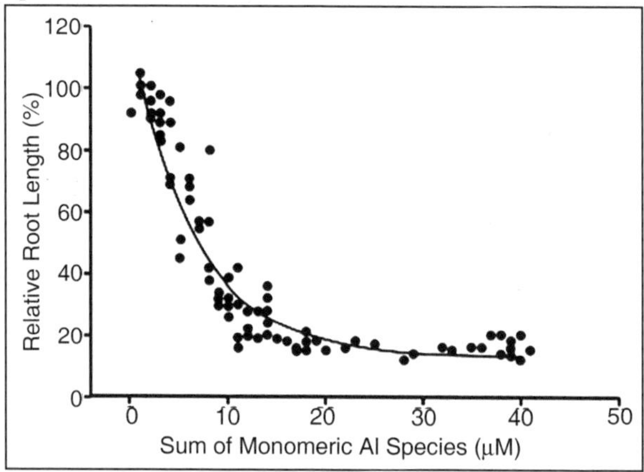


**Fig.** Soybean Root Growth Decreases Markedly as the Sum of Activities of Monomeric Al Species in Solution Increases.

Soluble Al in solution has a rapid effect on root growth, with the effects often visible within 2 d. Microscopically, decreased root growth may be evident within a few hours of exposure to Al, and changes in Golgi apparatus activity have been documented to occur within 5 h of placing roots in a solution containing Al. To be toxic, the root tip must be exposed to Al.

While the toxic effect of Al has been known for more than 60 years, the biochemical basis of Al toxicity has not been claried. However, Al is known to exert its primary toxic effect on roots, initial effects including a reduction in root length (through reduced cell elongation) and damage to cortical cells of the root epidermis near the root tip.

Proliferation of root hairs, important for uptake of water, essential nutrients and for infection by N-xing rhizobia in legumes, is severely reduced by Al at concentrations lower than those that reduce root growth. Proposals for the primary toxic effect have included reactions of Al with components of the cell wall, plasma membranes, cytoplasm and nucleus.

Visible effects of soluble Al on plant tops are considered to be a secondary effect through reduced nutrient uptake. Symptoms include those similar to deciencies of Ca, Mg, Fe, and P, probably as a result of decreased root proliferation and of reduced root activity. Further secondary effects include a reduction in uptake of water, increased sensitivity to drought and decreased $N_2$ xation by nodulated legumes.

Plant species differ markedly in tolerance to soluble Al in the soil solution. As with toxicity, the biochemical basis of genetic tolerance of Al toxicity is not clearly understood. However, malic acid excretion by root apices of Al-tolerant wheat is known to be ve- to ten-fold greater than that excreted by Al-sensitive wheat. Malic acid presumably complexes soluble Al, making it less toxic, and that capacity for excretion co-segregates with Al tolerance in progeny from crosses between near-isogenic lines. By implication, a single major gene is probably responsible for an Al tolerance that is functionally linked to Al stimulation of malic acid excretion.

## MANGANESE TOXICITY

Mn, an essential element for plants, is about the tenth most abundant element in the earth's crust. Like Al, Mn chemistry in acid soils is complex, affected by both soil pH and soil redox potential. Mn occurs in rocks mostly in co-ordination with $O^{2+}$ and as the divalent $Mn^{2+}$ in soil solution and in natural waters.

Biotic and abiotic oxidation occurs readily so that Mn does not generally occur to excess except in acid soils or in waterlogged soils (assuming there are sufcient Mn-containing minerals). Mn toxicity may occur following soil sterilisation due to loss of microorganisms that normally oxidise $Mn^{2+}$. Al is a root toxin, whereas high Mn is mainly toxic to shoots (at very high Mn concentration, root growth may be affected directly).

Thus, unlike Al, which often has an immediate effect on root proliferation, Mn must accumulate in shoots to become toxic. Three types of symptoms result from Mn accumulation: dark-brown, necrotic spots on lower leaves, distortion of expanding leaves (possibly an induced Ca deciency) and chlorosis of young leaves. Such chlorosis is often interpreted as an Fe deciency due to reduced Fe uptake by roots due to excess Mn.

Considerable genetic differences in tolerance to high concentrations of plant-available Mn exist among plant species, and even among lines within a species. Such variation may occur via differences in root exclusion of excess Mn, complexation of Mn within roots, or shoot tolerance of high Mn. Sunflower, watermelon and cucumber all show shoot tolerance and even excrete Mn in a biologically inactive form around trichomes (hairs) on leaves and stems.

External Mn concentrations at which plant growth is reduced also vary greatly among plant species. In carefully controlled solution culture, the critical external concentration (*i.e.* the concentration required for 10 per cent reduction in plant dry mass) in the two most sensitive species, maize and wheat, was 1.4 $\mu$M Mn. In contrast, sunflower growth was only reduced with 65 $\mu$M Mn in solution. Likewise, the critical internal Mn concentrations varied also, from 200 mg $kg^{-1}$ in maize to 5300 mg $kg^{-1}$ in sunflower. Although cowpea and bean had a similar external critical concentration, cowpea was able to tolerate a much higher tissue Mn. A higher retention of Mn in roots of soybean enabled this species to tolerate a higher external Mn concentration than cotton.

## COUNTERING ADVERSE EFFECTS OF ACID SOILS

Acid soil problems have been traditionally corrected via application of agricultural lime ($CaCO_3$) or dolomite ($CaCO_3$ + $MgCO_3$). Such amelioration raises soil pH, reducing Al and Mn concentrations in the soil solution, as well as adding the essential nutrients Ca and Mg. Lime or dolomite is generally required to raise soil pH to a level at which Al or Mn toxic-ities no longer affect plant growth (often above pH 5.5).

Rates required vary considerably depending on soil type, with clay soils requiring higher rates to increase soil pH (greater buffering capacity referred to earlier in connection with acidication). Moreover, lime requirements for sustainable agriculture vary with farming systems.

For example, lime required (kg $CaCO_3$ $ha^{-1}$ $year^{-1}$) to balance net acid accumulation is about: 0.8 for wool, 6.0 for lamb and 7–20 for cereal enterprises. Even though cereal production is not greatly acidifying, Australia's average wheat crop of 15 million t would have removed alkalinity equivalent to 135000 t $CaCO_3$ as well as 6000 t Ca, 18 000 t Mg and 66 000 t K (along with 330 000 t N, 42 000 t P and 26 000 t S) from the land.

Most of the crop was used for human consumption either in Australian cities or overseas. There would be little return of these nutrients to their original elds. Application of lime in Australia falls far below that required to balance losses of alkalinity.

Annually, *c.* 0.5 million t of lime is used in Australia, but more like 2.25 million t are needed annually as prophylactic dressings. Overliming is a potential problem, especially on sandy soils, due to reduced availability of essential micronutrients, especially Zn, while root diseases such as 'Take all' of cereals are exacerbated.

Amelioration of some acid soils is possible via application of gypsum ($CaSO_4.2H_2O$) which provides a readily soluble form of Ca. Leaching soluble Ca applied as gypsum may improve root growth in acid subsoil horizons. Application of Mo is also benecial where this essential micronutrient is in short supply due to low pH. Finally, breeding and selection for crop tolerance to acid soil limitations has immediate application in agriculture, but an understanding of factors involved is crucial because tolerance to one factor limiting growth on acid soils (Al toxicity) does not imply tolerance to another such as Mn excess or Mo deciency.

## CONCLUDING REMARKS

Acid soil infertility is a major limitation to plant growth worldwide, especially in the humid tropics, subtropics and temperate zones. Infertility results primarily from toxicities of Al or Mn and from deciencies of Ca, Mg, K, P and Mo. While there are many soils that are naturally acid, farming practices and industrial processes contribute to increased soil acidication. These include use of N fertilizers, especially where they exacerbate the loss of basic cations through leaching and crop removal, and growth of legumes, which increase soil N status. Acid rain is important in industrialised regions.

Two complementary approaches may be used to overcome the adverse effects of soil acidity. The rst is amelioration of acid soils through application of lime or gypsum to arrest degradation occasioned by a loss of basic cations. Unless these basic cations are replaced, land degradation through acidication and nutrient decline is inevitable. The second is a quest for genotypes better adapted to acid soil conditions. Genetic ap-proaches alone simply buy time and will result in further soil acidication, but do offer a benecial trade-off: tolerant plants would draw water and nutrients from acid subsoils which are difcult, if not impossible, to ameliorate any other way.

# 5

# Forest Entomology

Entomology is both a basic and an applied science that deals with the study of insects and their relatives. Entomologists are at the cutting edge of scientific research in such areas as systematics, physiology, biological control, and integrated pest management. Applied entomology helps further develop and transfer the knowledge and understanding gained through basic research to those who can benefit directly. Employers include universities, government agencies, and private industry. There are ample opportunities to work with people. No matter what your interests, skills, and background, there's a place for you in entomology. Because of its diversity, entomology provides many choices and opportunities for those interested in nature and the sciences. Some entomologists work in the field, others work in the laboratory or classroom, still others find a niche in regulatory entomology or international activities. If you like to work with computers, there are jobs developing software to aid farmers, foresters, and others in predicting and managing pest outbreaks. Computer models help them to apply pest control measures most effectively, with the least cost and the greatest safety. If you enjoy chemistry or physiology, you can conduct research on pheromones or population genetics. If your interest is in life sciences, you can use recombinant-DNA technology to improve plant and animal resistance to insect attacks.

## INSECT

Insects are a class of living creatures within the arthropods that have a chitinous exoskeleton, a three-part body (head, thorax, and abdomen), three pairs of jointed legs, compound eyes, and two antennae. They are among the most diverse groups of animals on the planet, including more than a million described species and represent more than half of all known living organisms.

The number of extant species is estimated at between six and ten million, and potentially represent over 90 per cent of the differing metazoan life forms on Earth. Insects may be found in nearly all environments, although only a small number of species occur in the oceans, a habitat dominated by another arthropod group, the crustaceans.

The life cycles of insects vary but most hatch from eggs. Insect growth is constrained by the inelastic exoskeleton and development involves a series of molts. The immature stages can differ from the adults in structure, habit and habitat and can include a passive pupal stage in those groups that undergo complete metamorphosis. Insects that undergo incomplete metamorphosis lack a pupal stage and adults develop through a series of nymphal stages. The higher level relationship of the hexapoda is unclear. Fossilized insects of enormous size have been found from the Paleozoic Era, including giant dragonflies with wingspans of 55 to 70 cm (22–28 in). The most diverse insect groups appear to have coevolved with flowering plants.

Insects typically move about by walking, flying or occasionally swimming. As it allows for rapid yet stable movement, many insects adopt a tripedal gait in which they walk with their legs touching the ground in alternating triangles. Insects are the only invertebrates to have evolved flight. Many insects spend at least part of their life underwater, with larval adaptations that include gills and some adult insects are aquatic and have adaptations for swimming.

Some species, like water striders, are capable of walking on the surface of water. Insects are mostly solitary, but some insects, such as certain bees, ants, and termites are social and live in large, well-organized colonies. Some insects, like earwigs, show maternal care, guarding their eggs and young. Insects can communicate with each other in a variety of ways. Male moths can sense the pheromones of female moths over distances of many kilometers. Other species communicate with sounds: crickets stridulate, or rub their wings together, to attract a mate and repel other males. Lampyridae in the beetle order Coleoptera communicate with light.

Humans regard certain insects as pests and attempt to control them using insecticides and a host of other techniques. Some insects damage crops by feeding on sap, leaves or fruits, a few bite humans and livestock, alive and dead, to feed on blood and some are capable of transmitting diseases to humans, pets and livestock. Nevertheless, without insects to pollinate flowers, the human race would soon run out of food because many of the crop plants that we rely on would not be able to reproduce. Many other insects are considered ecologically beneficial as predators and a few provide direct economic benefit. Silkworms and bees have been used extensively by humans for the production of silk and honey, respectively.

## DISTRIBUTION AND DIVERSITY

Even though the true dimensions of species diversity remain uncertain, estimates are ranging from 1.4 to 1.8 million species. This probably represents less than 20 per cent of all species on Earth, and with only about 20,000 new species of all organisms being described each year, it seems that most species will remain undescribed for many years unless there is a rapid increase in species descriptions. About 850,000–1,000,000 of all described species are insects. Of the 30 or so orders of insects, four dominate in terms of numbers of described species, with an estimated 600,000–795,000 species include Coleoptera, Diptera, Hymenoptera, and Lepidoptera. There are almost as many named species of beetle as there are of all other insects added together, or all other non-insects (plants and animals).

**Table. Comparison of the Estimated Number of Species for Vertebrates and the Four Most Species Order**

| | Described Species | Average Description Rate (Spesies per year) | Publication Effort |
|---|---|---|---|
| Coleoptera | 300,000-400,000 | 2308 | 0.01 |
| Lepidoptera | 110,000-120,000 | 642 | 0.03 |
| Diptera | 90,000-150,000 | 1048 | 0.04 |
| Hymenoptera | 100,000-125,000 | 1196 | 0.02 |

## MORPHOLOGY AND PHYSIOLOGY

### External

Insects have segmented bodies supported by an exoskeleton, a hard outer covering made mostly of chitin. The segments of the body are organized into three distinctive but interconnected units, or tagmata: a head, a thorax, and an abdomen. The head supports a pair of sensory antennae, a pair of compound eyes, and, if present, one to three simple eyes (or ocelli) and three sets of variously modified appendages that form the mouthparts.

The thorax has six segmented legs—one pair each for the prothorax, mesothorax and the metathorax segments making up the thorax—and, if present in the species, two or four wings. The abdomen consists of eleven segments, though in a few species of insects these segments may be fused together or reduced in size. The abdomen also contains most of the digestive, respiratory, excretory and reproductive internal structures. There is considerable variation and many adaptations in the body parts of insects especially wings, legs, antenna, mouth-parts etc.

### Segmentation

The head: is enclosed in a hard, heavily sclerotized, unsegmented, exoskeletal head capsule, or epicranium, which contains most of the sensing organs, including the antennae, ocellus or eyes, and the mouthparts. Out of all the insect orders, Orthoptera displays the most features found in other insects, including the sutures and sclerites. here, the vertex, or the apex (dorsal region), is situated between the compound eyes for insects with a hypognathous and opisthognathous head. In prognathous insects, the vertex is not found between the compound eyes, but rather, where the ocelli are normally. This is because the primary axis of the head is rotated 90 degrees to become parallel to the primary axis of the body. In some species this region is modified and assumes a different name.

The thorax: is a segment composed of three sections, the prothorax, mesothorax, and the metathorax. The anterior segment, closest to the head, is the prothorax, with the major features being the first pair of legs and the pronotum. The middle segment is the mesothorax with the major features being the second pair of legs and the anterior wings. The third and most posterior segment, abutting the abdomen, is the metathorax, which features the third pair of legs and the posterior wings.

Each segment is dilineated by an intersegmental suture. Each segment has four basic regions. The dorsal surface is called the tergum (or notum) to distinguish them from the abdominal terga. The two lateral regions are called the pleura (singular: pleuron) and the ventral aspect is called the sternum. In turn, the notum of the prothorax is called the pronotum, the notum for the mesothorax is called the mesonotum and the notum for the metathorax is called the metanotum. Continuing with this logic, there is also the mesopleura and metapleura as well as the mesosternum and metasternum. The abdomen: is a the last segment of the insect, which typically consists of 11–12 segments and is less strongly sclerotized than the head or thorax. Each segment of the abdomen is represented by a sclerotized tergum, sternum, and perhaps a pleurite. Terga are separated from each other and from the adjacent sterna or pleura by a membrane. Spiracles are located in the pleural area.

Variation of this ground plan includes the fusion of terga or terga and sterna to form continuous dorsal or ventral shields or a conical tube. Some insects bear a sclerite in the pleural area called a laterotergite. Ventral sclerites are sometimes called laterosternites. During the embryonic stage of many insects and the postembryonic stage of primitive insects, 11 abdominal segments are present. In modern insects there is a tendency towards reduction in the number of the abdominal segments, but the primitive number of 11 is maintained during embryogenesis

Variation in abdominal segment number is considerable. If the Apterygota are considered to be indicative of the ground plan for pterygotes, confusion reigns: adult Protura have 12 segments, Collembola have 6. The orthopteran family Acrididae has 11 segments, and a fossil specimen of Zoraptera has a 10-segmented abdomen.

### *Exoskeleton*

Insect outer skeleton, the cuticle, is made up of two layers: the epicuticle, which is a thin and waxy water resistant outer layer and contains no chitin, and a lower layer called the procuticle. The procuticle is chitinous and much thicker than the epicuticle and has two layers: an outer layer known as the exocuticle and an inner layer known as the endocuticle. The tough and flexible endocuticle is built from numerous layers of fibrous chitin and proteins, criss-crossing each others in a sandwich pattern, while the exocuticle is rigid and hardened. The exocuticle is greatly reduced in many soft-bodied insects (*e.g.,* caterpillars), especially during their larval stages.

Insects are the only invertebrates to have developed active flight capability, and this has played an important role in their success. Their muscles are able to contract multiple times for each single nerve impulse, allowing the wings to beat faster than would ordinarily be possible.

Having their muscles attached to their exoskeletons is more efficient and allows more muscle connections; crustaceans also use the same method, though all spiders use hydraulic pressure to extend their legs, a system inherited from their pre-arthropod ancestors. Unlike insects, though, most aquatic crustaceans are biomineralized with calcium carbonate extracted from the water.

**Internal**

***Nervous system***

The nervous system of an insect can be divided into a brain and a ventral nerve cord. The head capsule is made up of six fused segments, each with a pair of ganglia, or a cluster of nerve cells outside of the brain. The first three pairs of ganglia are fused into the brain, while the three following pairs are fused into a structure of three pairs of ganglia under the insect's esophagus, called the subesophageal ganglion.

The thoracic segments have one ganglion on each side, which are connected into a pair, one pair per segment. This arrangement is also seen in the abdomen but only in the first eight segments. Many species of insects have reduced numbers of ganglia due to fusion or reduction. Some cockroaches have just six ganglia in the abdomen, whereas the wasp Vespa crabro has only two in the thorax and three in the abdomen. Some insects, like the house fly Musca domestica, have all the body ganglia fused into a single large thoracic ganglion.

At least a few insects have nociceptors, cells that detect and transmit sensations of pain. This was discovered in 2003 by studying the variation in reactions of larvae of the common fruitfly Drosophila to the touch of a heated probe and an unheated one.

The larvae reacted to the touch of the heated probe with a stereotypical rolling behaviour that was not exhibited when the larvae were touched by the unheated probe. Although nociception has been demonstrated in insects, there is not a consensus that insects feel pain consciously but see Pain in invertebrates.

### Digestive System

An insect uses its digestive system to extract nutrients and other substances from the food it consumes. Most of this food is ingested in the form of macromolecules and other complex substances like proteins, polysaccharides, fats, and nucleic acids. These macromolecules must be broken down by catabolic reactions into smaller molecules like amino acids and simple sugars before being used by cells of the body for energy, growth, or reproduction. This break-down process is known as digestion. The main structure of an insect's digestive system is a long enclosed tube called the alimentary canal, which runs lengthwise through the body. The alimentary canal directs food unidirectionally from the mouth to the anus. It has three sections, each of which performs a different process of digestion. In addition to the alimentary canal, insects also have paired salivary glands and salivary reservoirs. These structures usually reside in the thorax, adjacent to the foregut.

The salivary glands (element 30 in numbered diagram) in an insect's mouth produce saliva. The salivary ducts lead from the glands to the reservoirs and then forward through the head to an opening called the salivarium, located behind the hypopharynx. By moving its mouthparts (element 32 in numbered diagram) the insect can mix its food with saliva. The mixture of saliva and food then travels through the salivary tubes into the mouth, where it begins to break down. Some insects, like flies, have extra-oral digestion. Insects using extra-oral digestion expel digestive enzymes onto their food to break it down.

This strategy allows insects to extract a significant proportion of the available nutrients from the food source. The gut is where almost all of insects' digestion takes place. It can be divided into the foregut, midgut and hindgut.

The first section of the alimentary canal is the foregut (element 27 in numbered diagram), or stomodaeum. The foregut is lined with a cuticular lining made of chitin and proteins as protection from tough food. The foregut includes the buccal cavity (mouth), pharynx, esophagus, and Crop and proventriculus (any part may be highly modified) which both store food and signify when to continue passing onward to the midgut.

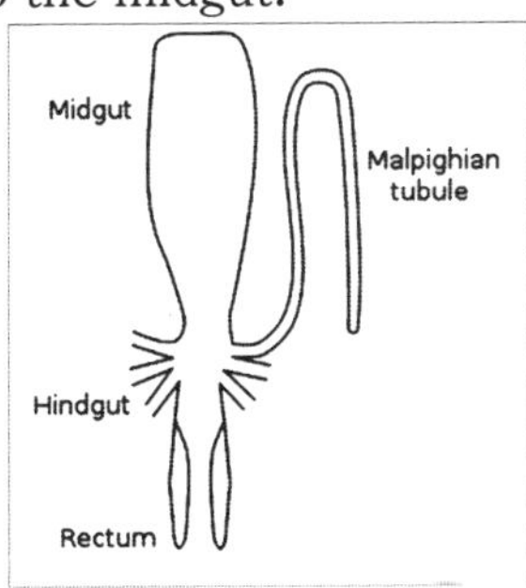


**Fig.** Stylized Diagram of Insect Digestive Tract Showing Malpighian Tubule, from an Insect of the Order Orthoptera.

Digestion starts in buccal cavity (mouth) as partially chewed food is broken down by saliva from the salivary glands. As the salivary glands produce fluid and carbohydrate-digesting enzymes (mostly amylases), strong muscles in the pharynx pump fluid into the buccal cavity, lubricating the food like the salivarium does, and helping blood feeders, and xylem and phloem feeders.

From there, the pharynx passes food to the esophagus, which could be just a simple tube passing it on to the crop and proventriculus, and then onward to the midgut, as in most insects. Alternately, the foregut may expand into a very enlarged crop and proventriculus, or the crop could just be a diverticulum, or fluid filled structure, as in some Diptera species.

### *Midgut*

Once food leaves the crop, it passes to the midgut (element 13 in numbered diagram), also known as the mesenteron, where the majority of digestion takes place. Microscopic projections from the midgut wall, called microvilli, increase the surface area of the wall and allow more nutrients to be absorbed; they tend to be close to the origin of the midgut. In some insects, the role of the microvilli and where they are located may vary. For example, specialized microvilli producing digestive enzymes may more likely be near the end of the midgut, and absorption near the origin or beginning of the midgut.

### *Hindgut*

In the hindgut (element 16 in numbered diagram), or proctodaeum, undigested food particles are joined by uric acid to form fecal pellets. The rectum absorbs 90 per cent of the water in these fecal pellets, and the dry pellet is then eliminated through the anus (element 17), completing the process of digestion. The uric acid is formed using hemolymph waste products diffused from the Malpighian tubules (element 20). It is then emptied directly into the alimentary canal, at the junction between the midgut and hindgut. The number of Malpighian tubules possessed by a given insect varies between species, ranging from only two tubules in some insects to over 100 tubules in others.

## Endocrine System

The salivary glands (element 30 in numbered diagram) in an insect's mouth produce saliva. The salivary ducts lead from the glands to the reservoirs and then forward through the head to an opening called the salivarium, located behind the hypopharynx. By moving its mouthparts (element 32 in numbered diagram) the insect can mix its food with saliva. The mixture of saliva and food then travels through the salivary tubes into the mouth, where it begins to break down. Some insects, like flies, have extra-oral digestion. Insects using extra-oral digestion expel digestive enzymes onto their food to break it down. This strategy allows insects to extract a significant proportion of the available nutrients from the food source.

## Reproductive System

The reproductive system of female insects consist of a pair of ovaries, accessory glands, one or more spermathecae, and ducts connecting these parts. The ovaries are made up of a number of egg tubes, called ovarioles, which vary in size and number by species. The number of eggs that the insect is able to make vary by the number of ovarioles with the rate that eggs can be develop being also influenced by ovariole design. Female insects are able make eggs, receive and store sperm, manipulate sperm from different males, and lay eggs. Accessory glands or glandular parts of the oviducts produce a variety of substances for sperm maintenance, transport, and fertilization, as well as for protection of eggs. They can produce glue and protective substances for coating eggs or tough coverings for a batch of eggs called oothecae. Spermathecae are tubes or sacs in which sperm can be stored between the time of mating and the time an egg is fertilized.

For males, the reproductive system is the testis, suspended in the body cavity by tracheae and the fat body. Most male insects have a pair of testes, inside of which are sperm tubes or follicles that are enclosed within a membranous sac. The follicles connect to the vas deferens by the vas efferens, and the two tubular vasa deferentia connect to a median ejaculatory duct that leads to the outside. A portion of the vas deferens is often enlarged to form the seminal vesicle, which stores the sperm before they are discharged into the female. The seminal vesicles have glandular linings that secrete nutrients for nourishment and maintenance of the sperm. The ejaculatory duct is derived from an invagination of the epidermal cells during development and, as a result, has a cuticular lining. The terminal portion of the ejaculatory duct may be sclerotized to form the intromittent organ, the aedeagus. The remainder of the male reproductive system is derived from embryonic mesoderm, except for the germ cells, or spermatogonia, which descend from the primordial pole cells very early during embryogenesis.

### *Respiratory and Circulatory Systems*

Insect respiration is accomplished without lungs. Instead, the insect respiratory system uses a system of internal tubes and sacs through which gases either diffuse or are actively pumped, delivering oxygen directly to tissues that need it via their trachea (element 8 in numbered diagram). Since, oxygen is delivered directly, the circulatory system is not used to carry oxygen, and is therefore greatly reduced. The insect circulatory system has no veins or arteries, and instead consists of little more than a single, perforated dorsal tube which pulses peristaltically. Towards the thorax, the dorsal tube (element 14) divides into chambers and acts like the insect's heart. The opposite end of the dorsal tube is like the aorta of the insect circulating the hemolymph, arthropods' fluid analog of blood, inside the body cavity. Air is taken in through openings on the sides of the abdomen called spiracles. There are many different patterns of gas exchange demonstrated by different groups of insects. Gas exchange patterns in insects can range from continuous and diffusive ventilation, to discontinuous gas exchange.

During continuous gas exchange, oxygen is taken in and carbon dioxide is released in a continuous cycle. In discontinuous gas exchange, however, the insect takes in oxygen while it is active and small amounts of carbon dioxide are released when the insect is at rest. Diffusive ventilation is simply a form of continuous gas exchange that occurs by diffusion rather than physically taking in the oxygen. Some species of insect that are submerged also have adaptations to aid in respiration. As larvae, many insects have gills that can extract oxygen dissolved in water, while others need to rise to the water surface to replenish air supplies which may be held or trapped in special structures.

## REPRODUCTION AND DEVELOPMENT

The majority of insects hatch from eggs. The fertilization and development takes place inside the egg, enclosed by a shell (chorion). Some species of insects, like the cockroach Blaptica dubia, as well as juvenile aphids and tsetse flies, are ovoviviparous. The eggs of ovoviviparous animals develop entirely inside the female, and then hatch immediately upon being laid. Some other species, such as those in the genus of cockroaches known as Diploptera, are viviparous, and thus gestate inside the mother and are born alive. Some insects, like parasitic wasps, show polyembryony, where a single fertilized egg divides into many and in some cases thousands of separate embryos.

Other developmental and reproductive variations include haplodiploidy, polymorphism, paedomorphosis or peramorphosis, sexual dimorphism, parthenogenesis and more rarely hermaphroditism. In haplodiploidy, which is a type of sex-determination system, the offspring's sex is determined by the number of sets of chromosomes an individual receives.

This system is typical in bees and wasps. Polymorphism is the where a species may have different morphs or forms, as in the oblong winged katydid, which has four different varieties: green, pink, and yellow or tan. Some insects may retain phenotypes that are normally only seen in juveniles; this is called paedomorphosis. In peramorphosis, an opposite sort of phenomenon, insects take on previously unseen traits after they have matured into adults. Many insects display sexual dimorphism, in which males and females have notably different appearances, such as the moth Orgyia recens as an exemplar of sexual dimorphism in insects.

Some insects use parthenogenesis, a process in which the female can reproduce and give birth without having the eggs fertilized by a male. Many aphids undergo a form of parthenogenesis, called cyclical parthenogenesis, in which they alternate between one or many generations of asexual and sexual reproduction. In summer, aphids are generally female and parthenogenetic; in the autumn, males may be produced for sexual reproduction. Other insects produced by parthenogenesis are bees, wasps, and ants, in which they spawn males. However, overall, most individuals are female, which are produced by fertilization. The males are haploid and the females are diploid. More rarely, some insects display hermaphroditism, in which a given individual has both male and female reproductive organs.

Insect life-histories show adaptations to withstand cold and dry conditions. Some temperate region insects are capable of activity during winter, while some others migrate to a warmer climate or go into a state of torpor. Still other insects have evolved mechanisms of diapause that allow eggs or pupae to survive these conditions.

## METAMORPHOSIS

Metamorphosis in insects is the biological process of development all insects must undergo. There are two forms of metamorphosis: incomplete metamorphosis and complete metamorphosis.

### Incomplete Metamorphosis

Insects that show hemimetabolism, or incomplete metamorphosis, change gradually by undergoing a series of molts. An insect molts when it outgrows its exoskeleton, which does not stretch and would otherwise restrict the insect's growth. The molting process begins as the insect's epidermis secretes a new epicuticle. After this new epicuticle is secreted, the epidermis releases a mixture of enzymes that digests the endocuticle and thus detaches the old cuticle. When this stage is complete, the insect makes its body swell by taking in a large quantity of water or air, which makes the old cuticle split along predefined weaknesses where the old exocuticle was thinnest. Other arthropods have a much different process and only molt; though must accommodate for the difference in exoskeleton structure and make up with other enzymes.

Immature insects that go through incomplete metamorphosis are called nymphs or in the case of dragonflies and damselflies as naiads. Nymphs are similar in form to the adult except for the presence of wings, which are not developed until adulthood. With each molt, nymphs grow larger and become more similar in appearance to adult insects.

### *Complete Metamorphosis*

Holometabolism, or complete metamorphosis, is where the insect changes all in four stages, an egg or embryo, a larva, a pupa, and the adult or imago. In these species, egg hatches to produce a larva, which is generally worm-like in form. This worm-like form can be one of several varieties: eruciform (caterpillar-like), scarabaeiform (grub-like), campodeiform (elongated, flattened, and active), elateriform (wireworm-like) or vermiform (maggot-like). The larva grows and eventually becomes a pupa, a stage marked by reduced movement and often sealed within a cocoon. There are three types of pupae: obtect, exarate or coarctate. Obtect pupae are compact, with the legs and other appendages enclosed. Exarate pupae have their legs and other appendages free and extended. Coarctate pupae develop inside the larval skin. Insects undergo considerable change in form during the pupal stage, and emerge as adults. Butterflies are a well known example of an insects that undergo complete metamorphosis, although most insects use this life cycle. Some insects have evolved this system to hypermetamorphosis.

Some of the oldest and most successful insect groups, such Endopterygota, use a system of complete metamorphosis. Strangely though, complete metamorphosis is unique to certain insect orders, like Diptera, Lepidoptera, and Hymenoptera, and no other arthropods undergo it, but incomplete metamorphosis.

**Senses and Communication**

Many insects possess very sensitive and/or specialized organs of perception. Some insects such as bees can perceive ultraviolet wavelengths, or detect polarized light, while the antennae of male moths can detect the pheromones of female moths over distances of many kilometers. There is a pronounced tendency for there to be a trade-off between visual acuity and chemical or tactile acuity, such that most insects with well-developed eyes have reduced or simple antennae, and vice-versa.

There are a variety of different mechanisms by which insects perceive sound, while the patterns are not universal, insects can generally hear sound if they can produce it. Different insect species can have varying hearing, though most insects can hear only a narrow range of frequencies related to the frequency of the sounds they can produce. Mosquitoes have been found to hear up to 2 MHz., and some grasshoppers can hear up to 50 MHz. Certain predatory and parasitic insects can detect the characteristic sounds made by their prey or hosts, respectively. For instance, some nocturnal moths can perceive the ultrasonic emissions of bats, which helps them avoid predation. Insects that feed on blood have special sensory structures that can detect infrared emissions, and use them to home in on their hosts.

Some insects display a rudimentary sense of numbers, such as the solitary wasps that prey upon a single species. The mother wasp lays her eggs in individual cells and provides each egg with a number of live caterpillars on which the young feed when hatched. Some species of wasp always provide five, others twelve, and others as high as twenty-four caterpillars per cell. The number of caterpillars is different among species, but always the same for each sex of larva. The male solitary wasp in the genus Eumenes is smaller than the female, so the mother of one species supplies him with only five caterpillars; the larger female receives ten caterpillars in her cell.

## Light Production and Vision

A few insects, such as members of the families Poduridae and Onychiuridae (Collembola), Mycetophilidae (Diptera), and the beetle families Lampyridae, Phengodidae, Elateridae and Staphylinidae are bioluminescent. The most familiar group are the fireflies, beetles of the family Lampyridae. Some species are able to control this light generation to produce flashes. The function varies with some species using them to attract mates, while others use them to lure prey. Cave dwelling larvae of Arachnocampa (Mycetophilidae, Fungus gnats) glow to lure small flying insects into sticky strands of silk. Some fireflies of the genus Photuris mimic the flashing of female Photinus species to attract males of that species, which are then captured and devoured. The colours of emitted light vary from dull blue (Orfelia fultoni, Mycetophilidae) to the familiar greens and the rare reds (Phrixothrix tiemanni, Phengodidae).

Most insects, except some species of cave crickets, are able to perceive light and dark. Many species have acute vision capable of detecting minute movements. The eyes include simple eyes or ocelli as well as compound eyes of varying sizes. Many species are able to detect light in the infrared, ultraviolet and the visible light wavelengths. Colour vision has been demonstrated in many species and phylogenetic analysis suggests that UV-green-blue trichromacy existed from at least the Devonian period between 416 and 359 million years ago.

## Sound Production and Hearing

Insects were the earliest organisms to produce and sense sounds. Insects make sounds mostly by mechanical action of appendages. In grasshoppers and crickets, this is achieved by stridulation. Cicadas make the loudest sounds among the insects by producing and amplifying sounds with special modifications to their body and musculature. The African cicada Brevisana brevis has been measured at 106.7 decibels at a distance of 50 cm (20 in). Some insects, such as the hawk moths and Hedylid butterflies, can hear ultrasound and take evasive action when they sense that they have been detected by bats. Some moths produce ultrasonic clicks that were once thought to have a role in jamming bat echolocation. The ultrasonic clicks were subsequently found to be produced mostly by unpalatable moths to warn bats, just as warning colorations are used against predators that hunt by sight. Some otherwise palatable moths have evolved to mimic these calls.

Very low sounds are also produced in various species of Coleoptera, Hymenoptera, Lepidoptera, Mantodea, and Neuroptera. These low sounds are simply the sounds made by the insect's movement. Through microscopic stridulatory structures located on the insect's muscles and joints, the normal sounds of the insect moving are amplified and can be used to warn or communicate with other insects. Most sound-making insects also have tympanal organs that can perceive airborne sounds. Some species in Hemiptera, such as the corixids (water boatmen), are known to communicate via underwater sounds. Most insects are also able to sense vibrations transmitted through surfaces. For example, an insect is caught in a spider web and struggles to escape. The vibrations it produces are sensed by the spider, who is alerted to its presence. Through these vibrations, the spider can tell where on the web the insect is located, as well as how big it is. Communication using surface-borne vibrational signals is more widespread among insects because of size constraints in producing air-borne sounds. Insects cannot effectively produce low-frequency sounds, and high-frequency sounds tend to disperse more in a dense environment (such as foliage), so insects living in such environments communicate primarily using substrate-borne vibrations. The mechanisms of production of vibrational signals are just as diverse as those for producing sound in insects.

Some species use vibrations for communicating within members of the same species, such as to attract mates as in the songs of the shield bug Nezara viridula. Vibrations can also be used to communicate between entirely different species; lycaenid (gossamer-winged butterfly) caterpillars which are myrmecophilous (living in a mutualistic association with ants) communicate with ants in this way. The Madagascar hissing cockroach has the ability to press air through its spiracles to make a hissing noise as a sign of aggression; the Death's-head Hawkmoth makes a squeaking noise by forcing air out of their pharynx when agitated, which may also reduce aggressive worker honey bee behaviour when the two are in close proximity.

**Chemical Communication**

In addition to the use of sound for communication, a wide range of insects have evolved chemical means for communication. These chemicals, termed semiochemicals, are often derived from plant metabolites include those meant to attract, repel and provide other kinds of information. Pheromones, a type of semiochemical, are used for attracting mates of the opposite sex, for aggregating conspecific individuals of both sexes, for deterring other individuals from approaching, to mark a trail, and to trigger aggression in nearby individuals. Allomonea benefit their producer by the effect they have upon the receiver. Kairomones benefit their receiver instead of their producer.

Synomones benefit the producer and the receiver. While some chemicals are targeted at individuals of the same species, others are used for communication across species. The use of scents is especially well known to have developed in social insects.

## Social Behaviour

Social insects, such as termites, ants and many bees and wasps, are the most familiar species of eusocial animal. They live together in large well-organized colonies that may be so tightly integrated and genetically similar that the colonies of some species are sometimes considered superorganisms. It is sometimes argued that the various species of honey bee are the only invertebrates (and indeed one of the few non-human groups) to have evolved a system of abstract symbolic communication where a behaviour is used to represent and convey specific information about something in the environment. In this communication system, called dance language, the angle at which a bee dances represents a direction relative to the sun, and the length of the dance represents the distance to be flown.

Only insects which live in nests or colonies demonstrate any true capacity for fine-scale spatial orientation or homing. This can allow an insect to return unerringly to a single hole a few millimeters in diameter among thousands of apparently identical holes clustered together, after a trip of up to several kilometers' distance. In a phenomenon known as philopatry, insects that hibernate have shown the ability to recall a specific location up to a year after last viewing the area of interest. A few insects seasonally migrate large distances between different geographic regions (*e.g.,* the overwintering areas of the Monarch butterfly).

## Care of Young

Some bees, ants, termites and some wasps are eusocial, build nest, guard eggs and provide food for offsprings fulltime. Most insects, however, lead short lives as adults, and rarely interact with one another except to mate or compete for mates. A small number exhibit some form of parental care, where they will at least guard their eggs, and sometimes continue guarding their offspring until adulthood, and possibly even feeding them. Another simple form of parental care is to construct a nest (a burrow or an actual construction, either of which may be simple or complex), store provisions in it, and lay an egg upon those provisions. The adult does not contact the growing offspring, but it nonetheless does provide food. This sort of care is typical for most species of bees and various types of wasps.

## Locomotion

### *Flight*

Insects are the only group of invertebrates to have developed flight. The evolution of insect wings has been a subject of debate. Some entomologists suggest that the wings are from paranotal lobes, or extensions from the insect's exoskeleton called the nota, called the paranotal theory. Other theories are based on a pleural origin. These theories include suggestions that wings originated from modified gills, spiracular flaps or as from an appendage of the epicoxa. The epicoxal theory suggests the insect wings are modified epicoxal exites, a modified appendage at the base of the legs or coxa. In the Carboniferous age, some of the Meganeura dragonflies had as much as a 50 cm (20 in) wide wingspan. The appearance of gigantic insects has been found to be consistent with high atmospheric oxygen. The respiratory system of insects constrains their size, however the high oxygen in the atmosphere allowed larger sizes. The largest flying insects today are much smaller and include several moth species such as the Atlas moth and the White Witch (Thysania agrippina). Insect flight has been a topic of great interest in aerodynamics due partly to the inability of steady-state theories to explain the lift generated by the tiny wings of insects.

Unlike birds, many small insects are swept along by the prevailing winds although many of the larger insects are known to make migrations. Aphids are known to be transported long distances by low-level jet streams. As such, fine line patterns associated with converging winds within weather radar imagery, like the WSR-88D radar network, often represent large groups of insects.

**Walking**

Many adult insects use six legs for walking and have adopted a tripedal gait. The tripedal gait allows for rapid walking while always having a stable stance and has been studied extensively in cockroaches. The legs are used in alternate triangles touching the ground. For the first step, the middle right leg and the front and rear left legs are in contact with the ground and move the insect forward, while the front and rear right leg and the middle left leg are lifted and moved forward to a new position. When they touch the ground to form a new stable triangle the other legs can be lifted and brought forward in turn and so on. The purest form of the tripedal gait is seen in insects moving at high speeds. However, this type of locomotion is not rigid and insects can adapt a variety of gaits. For example, when moving slowly, turning, or avoiding obstacles, four or more feet may be touching the ground. Insects can also adapt their gait to cope with the loss of one or more limbs.

Cockroaches are among the fastest insect runners and, at full speed, adopt a bipedal run to reach a high velocity in proportion to their body size. As cockroaches move very quickly, they need to be video recorded at several hundred frames per second to reveal their gait. More sedate locomotion is seen in the stick insects or walking sticks (Phasmatodea). A few insects have evolved to walk on the surface of the water, especially the bugs of the Gerridae family, commonly known as water striders. A few species of ocean-skaters in the genus Halobates even live on the surface of open oceans, a habitat that has few insect species.

## Use in Robotics

Insect walking is of particular interest as an alternative form of locomotion in robots. The study of insects and bipeds has a significant impact on possible robotic methods of transport. This may allow new robots to be designed that can traverse terrain that robots with wheels may be unable to handle.

### *Swimming*

A large number of insects live either parts or the whole of their lives underwater. In many of the more primitive orders of insect, the immature stages are spent in an aquatic environment. Some groups of insects, like certain water beetles, have aquatic adults as well.

Many of these species have adaptations to help in under-water locomotion. Water beetles and water bugs have legs adapted into paddle-like structures. Dragonfly naiads use jet propulsion, forcibly expelling water out of their rectal chamber. Some species like the water striders are capable of walking on the surface of water. They can do this because their claws are not at the tips of the legs as in most insects, but recessed in a special groove further up the leg; this prevents the claws from piercing the water's surface film. Other insects such as the Rove beetle Stenus are known to emit pygidial gland secretions that reduce surface tension making it possible for them to move on the surface of water by Marangoni propulsion (also known by the German term Entspannung-sschwimmen).

## Phylogeny and Systematic

The evolutionary relationships of insects to other animal groups remain unclear. Although more traditionally grouped with millipedes and centipedes, evidence has emerged favouring closer evolutionary ties with crustaceans. In the Pancrustacea theory, insects, together with Remipedia and Malacostraca, make up a natural clade. Other terrestrial arthropods, such as centipedes, millipedes, scorpions and spiders, are sometimes confused with insects since, their body plans can appear similar, sharing (as do all arthropods) a jointed exoskeleton.

However, upon closer examination their features differ significantly; most noticeably they do not have the six legs characteristic of adult insects. The higher-level phylogeny of the arthropods continues to be a matter of debate and research. In 2008, researchers at Tufts University uncovered what they believe is the world's oldest known full-body impression of a primitive flying insect, a 300 million-year-old specimen from the Carboniferous Period. The oldest definitive insect fossil is the Devonian Rhyniognatha hirsti, from the 396 million year old Rhynie chert. It may have superficially resembled a modern-day silverfish insect. This species already possessed dicondylic mandibles (two articulations in the mandible), a feature associated with winged insects, suggesting that wings may already have evolved at this time. Thus, the first insects probably appeared earlier, in the Silurian period.

There have been four super radiations of insects: beetles (evolved ~300 million years ago), flies (evolved ~250 million years ago), moths and wasps (evolved ~150 million years ago). These four groups account for the majority of described species. The flies and moths along with the fleas evolved from the Mecoptera.

The origins of insect flight remain obscure, since, the earliest winged insects currently known appear to have been capable fliers. Some extinct insects had an additional pair of winglets attaching to the first segment of the thorax, for a total of three pairs. As of 2009, there is no evidence that suggests that the insects were a particularly successful group of animals before they evolved to have wings.

Late Carboniferous and Early Permian insect orders include both extant groups and a number of Paleozoic species, now extinct. During this era, some giant dragonfly-like forms reached wingspans of 55 to 70 cm (22 to 28 in) making them far larger than any living insect. This gigantism may have been due to higher atmospheric oxygen levels that allowed increased respiratory efficiency relative to today. The lack of flying vertebrates could have been another factor. Most extinct orders of insects developed during the Permian period that began around 270 million years ago. Many of the early groups became extinct during the Permian-Triassic extinction event, the largest mass extinction in the history of the Earth, around 252 million years ago.

The remarkably successful Hymenopterans appeared as long as 146 million years ago in the Cretaceous period, but achieved their wide diversity more recently in the Cenozoic era, which began 66 million years ago. A number of highly successful insect groups evolved in conjunction with flowering plants, a powerful illustration of coevolution.

Many modern insect genera developed during the Cenozoic. Insects from this period on are often found preserved in amber, often in perfect condition. The body plan, or morphology, of such specimens is thus easily compared with modern species. The study of fossilized insects is called paleoentomology.

**Evolutionary Relationships**

Insects are prey for a variety of organisms, including terrestrial vertebrates. The earliest vertebrates on land existed 400 million years ago and were large amphibious piscivores, through gradual evolutionary change, insectivory was the next diet type to evolve.

Insects were among the earliest terrestrial herbivores and acted as major selection agents on plants. Plants evolved chemical defences against this herbivory and the insects in turn evolved mechanisms to deal with plant toxins. Many insects make use of these toxins to protect themselves from their predators. Such insects often advertise their toxicity using warning colours. This successful evolutionary pattern has also been utilized by mimics. Over time, this has led to complex groups of coevolved species. Conversely, some interactions between plants and insects, like pollination, are beneficial to both organisms. Coevolution has led to the development of very specific mutualisms in such systems.

**Taxonomy**

Traditional morphology-based or appearance-based systematics has usually given Hexapoda the rank of superclass, and identified four groups within it: insects (Ectognatha), springtails (Collembola), Protura and Diplura, the latter three being grouped together as Entognatha on the basis of internalized mouth parts. Supraordinal relationships have undergone numerous changes with the advent of methods based on evolutionary history and genetic data. A recent theory is that Hexapoda is polyphyletic (where the last common ancestor was not a member of the group), with the entognath classes having separate evolutionary histories from Insecta. Many of the traditional appearance-based taxa have been shown to be paraphyletic, so rather than using ranks like subclass, superorder and infraorder, it has proved better to use monophyletic groupings (in which the last common ancestor is a member of the group). The following represents the best supported monophyletic groupings for the Insecta.

Insects can be divided into two groups historically treated as subclasses: wingless insects, known as Apterygota, and winged insects, known as Pterygota. The Apterygota consist of the primitively wingless order of the silverfish (Thysanura). Archaeognatha make up the Monocondylia based on the shape of their mandibles, while Thysanura and Pterygota are grouped together as Dicondylia. It is possible that the Thysanura themselves are not monophyletic, with the family Lepidotrichidae being a sister group to the Dicondylia (Pterygota and the remaining Thysanura).

Paleoptera and Neo-ptera are the winged orders of insects differentiated by the presence of hardened body parts called sclerites; also, in Neo-ptera, muscles that allow their wings to fold flatly over the abdomen. Neo-ptera can further be divided into incomplete metamorphosis-based (Polyneoptera and Paraneoptera) and complete metamorphosis-based groups. It has proved difficult to clarify the relationships between the orders in Polyneoptera because of constant new findings calling for revision of the taxa. For example, Paraneoptera has turned out to be more closely related to Endopterygota than to the rest of the Exopterygota. The recent molecular finding that the traditional louse orders Mallophaga and Anoplura are derived from within Psocoptera has led to the new taxon Psocodea. Phasmatodea and Embiidina have been suggested to form Eukinolabia. Mantodea, Blattodea and Isoptera are thought to form a monophyletic group termed Dictyoptera.

It is likely that Exopterygota is paraphyletic in regard to Endopterygota. Matters that have had a lot of controversy include Strepsiptera and Diptera grouped together as Halteria based on a reduction of one of the wing pairs – a position not well-supported in the entomological community. The Neuropterida are often lumped or split on the whims of the taxonomist. Fleas are now thought to be closely related to boreid mecopterans. Many questions remain to be answered when it comes to basal relationships amongst endopterygote orders, particularly Hymenoptera.

The study of the classification or taxonomy of any insect is called systematic entomology. If one works with a more specific order or even a family, the term may also be made specific to that order or family, for example systematic dipterology.

## ECOLOGY

Insect ecology is the scientific study of how insects, individually or as a community, interact with the surrounding environment or ecosystem. Insects play one of the most important roles in their ecosystems, which includes many roles, such as dung burial, pest control, pollination, and wildlife nutrition. An example is beetles, which are scavengers that feed on dead animals and fallen trees and thereby recycle biological materials into forms found useful by other organisms. These insects, and others, are responsible for much of the process by which topsoil is created.

### Defence and Predation

Insects are mostly soft bodied, fragile and almost defenceless compared to other, larger lifeforms, while the immature stages are small, move slowly or are immobile, hence, all stages are exposed to predation and parasitism. So insects have a variety of strategies to avoid being attacked by predators or parasitoids. These include camouflage, mimicry, toxicity, and active defence.

Camouflage is also important defence strategies, which involves the use of coloration or shape to blend into the surrounding environment. This sort of protective coloration is common and widespread among beetle families, especially those that feed on wood or vegetation, such as many of the leaf beetles (family Chrysomelidae) or weevils. In some of these species, sculpturing or various coloured scales or hairs cause the beetle to resemble bird dung or other inedible objects.

Many of those that live in sandy environments blend in with the coloration of the substrate. Most phasmids are known for effectively replicating the forms of sticks and leaves, and the bodies of some species (such as O. macklotti and Palophus centaurus) are covered in mossy or lichenous outgrowths that supplement their disguise. Some species have the ability to change colour as their surroundings shift (B. scabrinota, T. californica). In a further behavioural adaptation to supplement crypsis, a number of species have been noted to perform a rocking motion where the body is swayed from side to side that is thought to reflect the movement of leaves or twigs swaying in the breeze. Another method by which stick insects avoid predation and resemble twigs is by feigning death (catalepsy), where the insect enters a motionless state that can be maintained for a long period. The nocturnal feeding habits of adults also aids Phasmatodea in remaining concealed from predators.

Another defence that often uses colour or shape to deceive potential enemies is mimicry. A number of longhorn beetles (family Cerambycidae) bear a striking resemblance to wasps, which helps them avoid predation even though the beetles are in fact harmless. Batesian and Müllerian mimicry complexes are commonly found in Lepidoptera. Genetic polymorphism and natural selection give rise to otherwise edible species (the mimic) gaining a survival advantage by resembling inedible species (the model).

Such a mimicry complex is referred to as Batesian and is most commonly known by the mimicry by the limenitidine Viceroy butterfly of the inedible danaine Monarch. Later research has discovered that the Viceroy is, in fact more toxic than the Monarch and this resemblance should be considered as a case of Müllerian mimicry. In Müllerian mimicry, inedible species, usually within a taxonomic order, find it advantageous to resemble each other so as to reduce the sampling rate by predators who need to learn about the insects' inedibility. Taxa from the toxic genus Heliconius form one of the most well known Müllerian complexes.

Chemical defence is another important defence found amongst species of Coleoptera, usually being advertised by bright colours, such as the Monarch butterfly. They obtain their toxicity by sequestering the chemicals from the plants they eat into their own tissues. Some Lepidoptera manufacture their own toxins.

Predators that eat poisonous butterflies and moths may become sick and vomit violently, learning not to eat those types of species; this is actually the basis of Müllerian mimicry. A predator who has previously eaten a poisonous lepidopteran may avoid other species with similar markings in the future, thus saving many other species as well. Chemical defence is another important defence found amongst species of Coleoptera, with some species releasing chemicals in the form of a spray with surprising accuracy, such as ground beetles (Carabidae), may spray chemicals from their abdomen to repel predators.

Many insects are considered pests by humans. Insects commonly regarded as pests include those that are parasitic (mosquitoes, lice, bed bugs), transmit diseases (mosquitoes, flies), damage structures (termites), or destroy agricultural goods (locusts, weevils). Many entomologists are involved in various forms of pest control, as in research for companies to produce insecticides, but increasingly relying on methods of biological pest control, or biocontrol.

Biocontrol uses one organism to reduce the population density of another organism—the pest—and is considered a key element of integrated pest management.

Despite the large amount of effort focused at controlling insects, human attempts to kill pests with insecticides can backfire. If used carelessly the poison can kill all kinds of organisms in the area, including insects' natural predators such as birds, mice, and other insectivores. The effects of DDT's use exemplifies how some insecticides can threaten wildlife beyond intended populations of pest insects.

**As Beneficial**

Although pest insects attract the most attention, many insects are beneficial to the environment and to humans. Some insects, like wasps, bees, butterflies, and ants, pollinate flowering plants. Pollination is a mutualistic relationship between plants and insects. As insects gather nectar from different plants of the same species, they also spread pollen from plants on which they have previously fed. This greatly increases plants' ability to cross-pollinate, which maintains and possibly even improves their evolutionary fitness. This ultimately affects humans since, ensuring healthy crops is critical to agriculture.

A serious environmental problem is the decline of populations of pollinator insects, and a number of species of insects are now cultured primarily for pollination management in order to have sufficient pollinators in the field, orchard or greenhouse at bloom time. Insects also produce useful substances such as honey, wax, lacquer and silk. Honey bees have been cultured by humans for thousands of years for honey, although contracting for crop pollination is becoming more significant for beekeepers. The silkworm has greatly affected human history, as silk-driven trade established relationships between China and the rest of the world.

Insectivorous insects, or insects which feed on other insects, are beneficial to humans because they eat insects that could cause damage to agriculture and human structures. For example, aphids feed on crops and cause problems for farmers, but ladybugs feed on aphids, and can be used as a means to get significantly reduce pest aphid populations. While birds are perhaps more visible predators of insects, insects themselves account for the vast majority of insect consumption. Without predators to keep them in check, insects can undergo almost unstoppable population explosions.

Insects are also used in medicine, for example fly larvae (maggots) were formerly used to treat wounds to prevent or stop gangrene, as they would only consume dead flesh. This treatment is finding modern usage in some hospitals. Recently insects have also gained attention as potential sources of drugs and other medicinal substances. Also adult insects, such as crickets, and insect larvae of various kinds are also commonly used as fishing bait.

**In research**

Insects play important roles in biological research. For example, because of its small size, short generation time and high fecundity, the common fruit fly Drosophila melanogaster is a model organism for studies in the genetics of higher eukaryotes. D. melanogaster has been an essential part of studies into principles like genetic linkage, interactions between genes, chromosomal genetics, development, behaviour, and evolution. Because genetic systems are well conserved among eukaryotes, understanding basic cellular processes like DNA replication or transcription in fruit flies can help to understand those processes in other eukaryotes, including humans. The genome of D. melanogaster was sequenced in 2000, reflecting the organism's important role in biological research.

## IMPORTANCE OF STUDY INSECTS

Insects provide a readily accessible resource for you to use in developing a better scientific understanding of the world around you. For more than 350 million years, insects have evolved and adapted to become the creatures we know today. Through the millennia, insects have become an essential part of every terrestrial and freshwater ecosystem.

They are the most numerous and diverse form of life on Earth. About one million species are known, and it is estimated that 10 million are undiscovered. Some insects are easy to maintain in a laboratory or a classroom, and their short life cycles make them excellent subjects for research or teaching. The study of insects also helps us understand the physiology and biology of other animals. The study of their actions helps us understand animal social organization and behaviour. The study of insect populations helps us understand ecological interactions. The quantity and quality of insect life in and around a pond or stream, for example, can indicate the presence or absence of pollution. The study of insects helps us increase the bounty of the land and preserve its natural beauty. It also helps us understand how to protect lives and property from harmful insects.

**To Feed a Hungry World**

American agriculture is the world's leading producer and exporter of food. But American farmers must become even more efficient to compete in the world market. Food shortages still exist in many parts of the world. About 40 per cent of the world's food production is lost to insect pests each year. Sound entomological research and extension programmes are at the forefront of the sciences involving these important problems. Although reducing vast insect-caused losses will not automatically solve the hunger problems - other economic and cultural factors are important too - entomology is a central part of the solution.

**To Preserve the Diversity of Life**

Insects add to the natural beauty around us and are an integral part of the ecological web. Throughout the world habitat, alteration has caused the extinction of many organisms, including insects. By identifying endangered species and studying their habitats, entomologists help describe and restore threatened ecosystems. Entomologists are working to protect the environment and to sustain agricultural production, from the San Joaquin Valley to the Adirondack Mountains, and from the Sudan's parched grasslands to deep in Brazil's Amazon jungle.

In fact, tropical rain forests present special problems and opportunities for entomologists. Rain forests are among the oldest and most complex terrestrial ecosystems, but they are rapidly disappearing. Perhaps half of all plant and animal species are found only in tropical rain forests. Of the millions of insect species that remain to be discovered, most will be found in these forests. Their genetic diversity will include many beneficial species. Entomologists must discover these species before they cease to exist, and they must find ways to preserve the insects' habitats for future generations. Entomologists also help unravel the complexities of agricultural systems such as crops, livestock, forests, fishponds, and urban gardens.

Entomological research has helped the United States to become a model for all industrialized countries in solving certain public health problems. A century ago, malaria was a major problem in North America. It is now of minor importance. Our knowledge of the causes and means of controlling epidemics of insect-borne diseases has increased because of the hard work and dedication of entomologists. However, much work remains to be done. Few problems present greater challenges to medical and veterinary entomologists than the widespread distribution of insect-borne disease agents.

Vast areas of the world are dominated by insects that transmit parasites that cause yellow fever, river blindness, Chagas disease, and sleeping sickness. Malaria, plague, and tick-borne fevers are diseases of worldwide importance, and entomologists lead the way in research to combat these ailments. Health losses to insects are figured not only in lives but also in money. In North America alone, it costs hundreds of millions of dollars each year to control flies, grubs, lice, and ticks on livestock and poultry. Entomologists are seeking new, less expensive ways to prevent these losses. There is literally a world of opportunity awaiting you in the science of entomology. Whether your interests lie in conducting genetic research, developing computer software, exploring a remote rainforest, or helping to feed the world, an exciting and rewarding career awaits you.

## FRESHWATER UNIONOID MUSSELS

The lamellar layer of the shell is nacreous (pearly). The apertures are poorly developed and the shells are equivalve. Modern classifications of unionoid mussels recognize two families in North America; Margaritiferidae and Unionidae. Two subfamilies of unionids, Ambleminae and Unioninae, are recognized. Most amblemines are North American where they occur exclusively (or almost so) in the Atlantic and Gulf of Mexico drainages. Approximately 75 per cent of North American genera and 80 per cent of species belong to Ambleminae. The North American Ambleminae belong to the tribes Amblemini, Pleurobemini, and Lampsilini.

Native North American freshwater bivalves are either mussels in the order Unionida (Margaritiferidae and Unionidae) or fingernail clams (Veneroida: (Corbiculoidea: Sphaeriidae). Introduced bivalves include the now widely dispersed Asian clam, *Corbicula fluminea*, and the rapidly dispersing zebra mussel, *Dreissena polymorpha*.

Freshwater mussels (Unionoidea) occur worldwide but they are most abundant in North America, with their greatest diversity in the Ohio River Drainage of the Southeastern United States. They should not be confused with the unrelated marine mussels (Filibranchia, Pteriomorpha) from which they differ in many important respects. Freshwater mussels are usually found in medium to large rivers in water less than two meters deep.

A few species inhabit the quiet water of lakes but most are riverine. Small creeks and small lakes have few mussels although they may have diverse faunas of other bivalves such as the sphaeriid clams. Mussels live in soft, sedimentary bottoms with the foot and anterior end of the shell and body dug into the sediment with the posterior end exposed. Adult mussels move slowly and never very far from the site of settling after leaving the fish host. Dispersion occurs as larvae parasitic on fishes. Like other bivalves, mussels are suspension feeders using the gills as filters to remove particulate organic matter, detritus and plankton, from the water. Adults unionoids can be 4-30 cm in length.

Most freshwater mussels are gonochoric and cross fertilize without copulation. Fertilization is external and occurs in the water tubes of the female demibranchs where embryos are brooded to the larval stage. This is sometimes referred to as internal fertilization but, since the water tubes are part of the mantle cavity, and not the gonoduct, it is topologically outside the body and it is probably preferable to consider it to be external. The larva is the unique glochidium, an external parasite of fishes. Glochidia live only attached to the fish host and without the appropriate fish the life cycle cannot be completed.

Our freshwater mussels are imperiled by several human activities including habitat destruction (especially the impoundment of rivers), siltation from agricultural and construction run-off, organic and chemical pollution, competition with introduced exotics such as Asian clams and zebra mussels, extirpation of fish hosts for the larvae, and direct exploitation, originally for the button industry and now as a source of beadstock for the cultured pearl industry. Of about 300 unionoid taxa once known from North America 70 per cent are Endangered, Threatened, or of Special Concern. Twenty one taxa are Endangered and possibly Extinct, 77 are Endangered and extant, 43 are Threatened, 72 are of Special Concern, 14 are Undetermined, and only 70 taxa, or 24 per cent, are Currently Stable. Harvesting of mussels is not federally regulated except for species on the federal list of endangered or threatened species although many states regulate species and sizes that can be harvested.

Unionoid genera differ from each other in hinge morphology, hinge tooth morphology and development, shell thickness, the shell inflation, shell ornamentation, elaborations of the mantle margin, and use of the demibranchs for brooding. Species and genera are difficult to identify.

## LABORATORY SPECIMENS

Many native North American freshwater mussels are threatened or endangered and are protected by state and federal legislation. Obviously, these species should never be used for dissection in teaching laboratories. Whenever possible alternatives to native unionoid mussels should be used for the study of bivalve anatomy. Clams (*Mercenaria mercenaria*), marine mussels (*Mytilus edulis*), and oysters (*Crassostrea virginica*) are often available alive and inexpensively from local fish markets and supermarkets, even in the interior of the continent. If locally collected living specimens of freshwater bivalves are desired, it is much better to use the abundant, widespread, invasive Asian clam, *Corbicula fluminea*.

To avoid further damage to endangered and threatened populations native mussels should not be collected locally for laboratory use unless the instructor can identify the specimens to species. Federal legislations provide for substantial penalties for collection or possession of Threatened or Endangered species and some states require permits. Biological supply companies provide specimens of what we assume are species without conservation concern, classified as Currently Stable, for laboratory use. The actual identity of commercially supplied specimens varies and can only be known by identifying the specimens upon arrival at your institution. The catalogs advertise them as *Unio* or *Anodonta* but these names are used *sensu lato*, in the broad sense. Both genera, once large, have been split into several smaller genera and as the genera are currently defined, there are no species of *Unio* in the Americas.

*Anodonta* has thin shells without hinge teeth. Other genera, including genera such as *Actinonaias* and *Lampsilis* have thick shells and hinge teeth. Both types are marketed under the designation " *Anodonta* or *Unio*". This exercise was prepared using preserved, commercially supplied *Actinonaias* (marketed under the name " *Anodonta-Unio*") but can be used with fresh or preserved specimens of any unionoid mussel. The dissection is best conducted with the specimen covered with tap water although this may prove impracticable until both valves have been removed. A dissecting microscope should be used as needed. Although the exercise is written to be used with preserved specimens, it can also be used with living or freshly killed animals. Living specimens are very difficult to open due to the ability of the adductor muscles to hold the valves tightly together indefinitely. Literature accounts suggest placing living mussels in warm tap water (55-60 °C) for a few minutes. This is said to result in relaxation of the muscles and consequent gaping of the valves. Alternatively, mussels can be boiled or steamed briefly, until the valves gape open. Any procedure that kills the specimen is, of course, incompatible with later observation of ciliary activity on the gills, labial palps, and stomach walls or beating of the heart. Once the valves have separated, a sharp scalpel can be inserted between the valves to sever the anterior and posterior adductor muscles. Living specimens should be anesthetized in 5 per cent non-denatured ethanol in water.

## MOLLUSCA

Mollusca, the second largest metazoan taxon, consists of Aplacophora, Polyplacophora, Monoplacophora, Gastropoda, Cephalopoda, Bivalvia, and Scaphopoda. The typical mollusc has a calcareous shell, muscular foot, head with mouth and sense organs, and a visceral mass containing most of the gut, the heart, gonads, and kidney. Dorsally the body wall is the mantle and a fold of this body wall forms and encloses that all important molluscan chamber, the mantle cavity. The mantle cavity is filled with water or air and in it are located the gill(s), anus, nephridiopore(s) and gonopore(s). The coelom is reduced to small spaces including the pericardial cavity containing the heart and the gonocoel containing the gonad. The well-developed hemal system consists of the heart and vessels leading to a spacious hemocoel in which most of the viscera are located. The kidneys are large metanephridia. The central nervous system is cephalized and tetraneurous. There is a tendency to concentrate ganglia in the circumenteric nerve ring from which arise four major longitudinal nerve cords. Molluscs may be either gonochoric or hermaphroditic. Spiral cleavage produces a veliger larva in many taxa unless it is suppressed in favour of direct development or another larva. Molluscs arose in the sea and most remain there but molluscs have also colonized freshwater and terrestrial habitats.

### EUMOLLUSCA

Eumollusca, the sister taxon of Aplacophora, includes all molluscs other than aplacophorans. The eumolluscan gut has digestive ceca which are lacking in aplacophorans, the gut is coiled, and a complex radular musculature is present.

### CONCHIFERA

Conchifera, the sister taxon of Polyplacophora, includes all Recent molluscs other than aplacophorans and chitons. The conchiferan shell consists of an outer proteinaceous periostracum underlain by calcareous layers and is a single piece (although in some it may appear to be divided into two valves). The mantle margins are divided into three folds.

**GANGLIONEURA**

Most Recent molluscs are ganglioneurans, only the small taxa Aplacophora, Polyplacophora, and Monoplacophora are excluded. Neuron cell bodies are localized in ganglia.

**ANCYROPODA**

The mantle cavity, with its gills, is lateral. The calcareous portion of the shell is bivalve, with the valves opening laterally and joined dorsally by a derivative of the periostracum.

**BIVALVIA**

Bivalvia is a large, successful, and derived taxon. The body is laterally compressed and enclosed in a bivalve shell. The two valves are hinged dorsally. The the foot is large and adapted for digging in the ancestral condition. A crystalline style is usually present but never is there a radula. The mantle cavity is lateral and in most bivalves the gills are large and function in respiration and filter-feeding. The head is reduced and bears no special sense organs. The nervous system is not cephalized. The group includes scallops, clams, shipworms, coquinas, marine and freshwater mussels, oysters, cockles, zebra mussels, and many, many more.

**METABRANCHIA**

Metabranch gills are adapted for filter feeding. Water enters the mantle cavity posteriorly.

**EULAMELLIBRANCHIA**

Eulamellibranchs have gills with tissue interfilamentar connections.

## EXTERNAL ANATOMY

### SHELL

#### External Shell Features

Examine a cleaned dry shell from which the animal has been removed or, if such a shell is not available, an intact specimen with the animal still within the shell. The bivalve shell consists of two halves, or valves. The plane of symmetry passes between the two valves and divides the animal into right and left sides, which in most cases are more or less mirror images of each other.

The two valves articulate with each other along the dorsal midline via a hinge mechanism which may be simple and toothless (*Anodonta*, *Pyganodon*) or complex and toothed (most unionoids). The valves are held together along the hinge by a dark brown proteinaceous hinge ligament.

The valves can move laterally, away from each other, along the anterior, ventral, and posterior margins but are always in contact along the dorsal hinge. This lateral motion opens a space, known as the gape, along the anterior, ventral, and posterior margins and thus providing the animal with access to its environment. If you have an intact preserved specimen the gape will be present and obvious, although it will probably be larger than it would be in life. If you have a living specimen the valves will be held tightly together and there will be no gape.

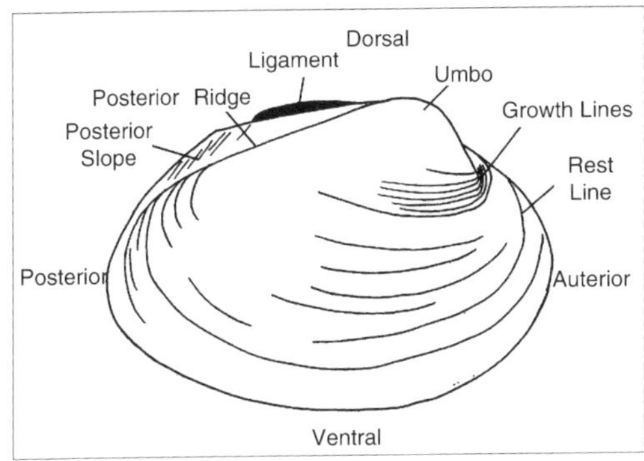


**Fig.** External view of the Right Valve of *Actinonaias*.

On each side of the hinge, anterior to the ligament, each valve displays a raised area known as theumbo, or beak (plural = umbones). The umbo is the oldest part of the shell. The two umbos almost touch across the midline. The umbo is anterior to the externally visible portion of the hinge ligament.

The anterior position of the umbo and its location with respect to the ligament can be used to recognize the anterior end of the clam without resorting to soft anatomy. Use the umbo to identify the anterior end of your mussel and the hinge to identify dorsal. Use these landmarks to find ventral and posterior and to determine which valve is theright valve and which is the left valve.

Relocate the ligament and note that it is outside the hinge and posterior to the umbo. In all freshwater bivalves the ligament is external. The posterior portion is obviously external and is clearly visible on the outside of the hinge even when the valves are closed. The anterior portion, although still external, is hidden when the valves are closed. The critical criterion being the position of the ligament with respect to the hinge teeth. If the ligament is outside the line of hinge teeth, it is external, if inside the line of teeth, it is considered to be internal. The ligament, being elastic, antagonizes the two adductor muscles and opens (abducts) the valves when the adductors are relaxed. An external ligament is stretched when the valves are closed and pulls the valves apart when the adductors are relaxed. An internal ligament, on the other hand, is compressed when the valves are closed and pushes the valves apart when the adductors relax.

The two valves of a unionoid mussel are similar to each other, a condition known as equivalve. Some clams, such as the marine oysters and jingles, have very different right and left valves and are inequivalve. Furthermore, the anterior and posterior ends of freshwater mussels are more or less similar to each other and the valves are said to be equilateral. Equilateral valves are symmetrical on a dorsal ventral axis. In contrast, the two ends of some bivalves, such as the marine mussels and pen clams are very different, are not symmetrical, and are inequilateral although they are equivalve.

Numerous closely spaced, fine, concentric growth lines are visible on the outside surface of the valves. Larger, more widely spaced, concentric ridges are rest lines. An oblique posterior ridge extends from the umbo in a posteroventral direction to the posterior margin of each valve. In *Actinonaias* the ridge is inconspicuous but in some genera it is strong and well developed. The region of the valve posterior to the ridge is the posterior slope. Depending on species, a variety of spines, corrugations, rays, or pustules may ornament the valves externally but none of these is present in *Actinonaias*. If you are using cleaned valves to study, go directly to "Internal Shell Features". If you have an intact specimen, skip to "Preview of Soft Anatomy".

**Internal Shell Features**

If you have a cleaned shell, its internal features can be studied now. If you have an intact specimen, you must wait until the shell has been opened and the animal removed before you can study the inside of the valves.

Relocate the hinge region along the dorsal margin of the valve. Find anterior and posterior, dorsal and ventral.

The inside surface of each valve bears scars of the muscles that attached to it. Among these are the conspicuous anterior and posterior adductor muscle scars. The anterior scar is the smaller of the two. The pallial line, along which the pallial muscles of the mantle insert, is a conspicuous line or groove paralleling the margin of the valve from the anterior adductor muscle scar to the posterior adductor muscle scar. The mantle is attached to the valve along this line. The small pedal protractor muscle scar is located near the posterior ventral margin of the anterior adductor scar. Dorsal to the protractor scar are the anterior pedal retractor muscle scars. The posterior pedal retractor muscle scar is dorsal to the posterior adductor scar and may be continuous with it or slightly separated from it. These muscles originate on the shell and insert on the foot. Their action is to help extend the foot out of the gape (pedal protractors) or withdraw the foot into the mantle cavity (pedal retractors). The umbo rises above the hinge and the ligament extends posteriorly from the umbo. Most unionoid hinges are equipped with hinge teeth to prevent shearing and keep the valves aligned when the powerful adductor muscles pull them together. The teeth are the pseudocardinal teeth and the lateral teeth. The teeth of the right and left valves are not symmetrical. Instead a tooth on one valve opposes, and fits snugly into, a depression on the opposite valve. The jagged, massive pseudocardinal teeth are far anterior, near the anterior adductor muscle scar. *Actinonaias*, like most unionoids has one large pseudocardinal tooth in the right hinge and two in the left. The lateral teeth are on the posterior end of the hinge and are long smooth ridges. The right hinge has one lateral teeth and the left hinge has two. The pesudocardinal teeth look a little like vertebrate teeth but the laterals do not. Compare the left and right valves to demonstrate to yourself the way the teeth fit together and prevent shearing.

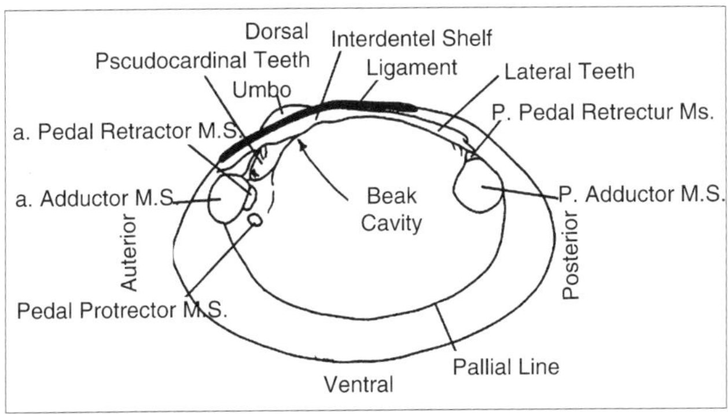


**Fig**. Internal view of a Right Valve of Actinonaias.
a. = Anterior, p. = Posterior, m.s. = Muscle Scar.

Anterior lateral teeth are absent in mussels and some lack teeth altogether (*viz.* Anodontinae; Anodonta, Pyganodon). The Asian clam, Corbicula, and the sphaeriid clams (both CorbiculoideaSF) have true cardinal teeth preceded and followed by anterior and posterior lateral teeth, respectively. In unionoids true cardinal teeth are absent and the ancestral anterior laterals have been modified to form the pseudocardinals.

Mollusc shells are secreted in layers by the mantle epithelium. The three layers of the typical bivalve shell are present and well developed in mussels. All three layers are secreted by the mantle, each by a specific region of secretory epithelium.

The outermost layer is the proteinaceous periostracum. In unionoid mussels its colour varies but is usually black, brown, or yellow. It may be patterned, often with rays. In older individuals thc periostracum is often damaged or eroded and much of it may be absent, especially near the umbo.

The innermost layer of the shell, the one adjacent to the animal itself, is the lamellar layer (= hypostracum) composed of thin calcareous sheets, or lamellae, of nacre (pronounced NAKE ur), or mother of pearl, layered on top of each other. The shiny lustrous nacre is visible on the inside of the valve. Most of the inside surface of the valve, except for the extreme outer border, is covered by nacre. The colour of the nacre varies with species.

The calcareous, white prismatic layer (= ostracum) lies between the periostracum and lamellar layers. It can be seen externally in areas near the umbo where the periostracum is eroded, exposing the white prismatic layer beneath it. It is composed of vertical prisms, polygonal in cross section.

The periostracum is impermeable to water and its presence protects the underlying calcium carbonate layers from dissolving and being eroded by the surrounding water. Its presence at the edge of the shell is essential for the continued secretion of new prismatic layer. It seals the margin and creates a chemically regulated extrapallial space in which new shell can be precipitated. In older areas of the shell, *i.e.* nearer the umbo, the protective periostracum is often broached and the prismatic layer may be exposed and subsequently eroded.

With the dissecting microscope examine a broken shell of a mussel, or other large bivalve such as *Mercenaria*, and view the shell layers in cross section. The piece you examine should include the original edge of the shell. The periostracum forms a thin organic layer on the outer surface. The prismatic and lamellar layers are distinctly different in this view. The prismatic layer, which lies immediately below the periostracum, is composed of conspicuous, more or less vertical, closely spaced columns, or prisms, of calcium carbonate. As you know, the prisms are vertical to the surface of the shell. The lamellar layer on the other hand consists of more or less horizontal sheets lying inside the prismatic layer. These lamellae are parallel to the surface of the shell. Note that the lamellar layer extends almost to the edge of the shell but that closest to the edge only the prismatic layer is present. The extreme edge is peripheral to the area where the lamellar layer is being secreted and consists entirely of recently secreted periostracum and prismatic layer with no lamellar layer yet present.

## PREVIEW OF SOFT ANATOMY

Commercially supplied preserved specimens will arrive gaping, usually with a wooden peg holding the valves apart in a wide ventral gape. The major features of the external soft anatomy can be previewed through this gape before a valve is removed. With living specimens this preview must wait until the adductor muscles have been cut. Look into the gape and note the most obvious features for use later as landmarks. You may have to force the valves a little farther apart to improve your view but, if so, be careful that you do not tear or otherwise damage the soft tissues.

The mantle, a molluscan apomorphy, is the dorsal body wall which typically exhibits modification in different ways in different classes. Under each of the two valves of a bivalve the mantle forms a thin sheet of soft tissue. These two sheets are the right and left mantle skirts (also known as mantle lobes or mantle folds by other authors) and they enclose the animal like a cloak (*i.e.* mantle). The two skirts are joined to each other and to the body along the dorsal midline, under the hinge. Laterally and ventrally the space enclosed by the two skirts is the mantle cavity. In life it is filled with circulating water. The circulation is driven by cilia on the gills. Posteriorly the margins of the right and left mantle skirts conspire to form a pair of openings, the ventral inhalant aperture and the dorsal exhalant aperture. Species of Unionidae and Margaritiferidae have simple apertures rather than siphons. Among the Unionoida only Mycetopodidae, Iridinidae, and Hyriidae have siphons.

In the center of the mantle cavity is the laterally compressed, anteriorly directed foot, which in preserved or dissected living specimens will be strongly contracted with a wavy ventral edge. The large mass of tissue dorsal to the foot is the visceral mass in which are located most of the organ systems including gonad, kidney, heart, and gut.

The bivalve head is weakly developed and lacks the concentration of sensory equipment typical of the heads of other molluscs. In bivalves most sensory receptors are located in the mantle margin, especially in the vicinity of the apertures.

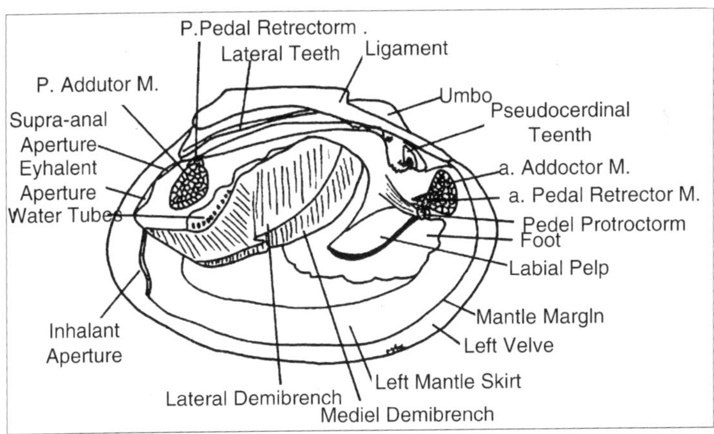


**Fig.** View of the External Soft Anatomy of the Right Side of an Undissected Specimen.

The right valve and mantle skirt have been removed resulting in the partial opening of the right exhalant chamber and exposure of a few water tubes belonging to the lateral demibranch. Because this is a preserved specimen tissues are unnaturally contracted. a = anterior, m = muscle, p = posterior. On each side of the visceral mass, between the mass and a mantle skirt, is a long, leaf-like gill. It may seem to you that there are two gills on each side but there is actually only one. At the anterior end of each gill there is a smaller, but also leaf-like labial palp. The palps resemble miniature gills and like them are double but there is actually only one on each side. Immediately anterior to the foot and labial palps is the anterior adductor muscle extending transversely from valve to valve. Hold the specimen so you can look into the posterior end of the mantle cavity.

Here you will see the posterior adductor muscle extending across the cavity from valve to valve. The action of the two adductor muscles is to bring the two valves together (adduct the valves).

## SOFT EXTERNAL ANATOMY

After previewing the soft anatomy remove the right valve as follows. Identify the right valve and then look into the gape and find the anterior and posterior adductor muscles. With a scalpel or blunt probe scrape the two adductor muscles away from their attachment to the right valve. This will be easy to do in preserved specimens, in fact, the adductors may already be separated from their attachments. This process should also separate the protractor and retractor muscles from the valve. (The adductors of fresh specimens must be cut with a sharp scalpel. The posterior muscle should be cut immediately to the right of the rectum. Be careful that you do not cut any tissues other than the two muscles.)

Separate the margin of the right mantle skirt from the periostracum that attaches it to the right valve. Gently lift the right valve away from the left, leaving the right mantle skirt behind, with the remainder of the mussel. Use the blunt, flat handle of the scalpel as necessary to push, not cut, the mantle skirt away from the inner surface of the right valve. You will notice that the right mantle skirt is attached to the right valve, not only by the periostracum along its edge, but also by a line of pallial (pallial = mantle) muscles that parallels the valve margin. As you break the connection of the pallial muscles with the valve note that a distinct pallial line marking the former attachment site is thereby revealed. All soft tissues, including the right mantle skirt, should remain cradled in the left valve. Break or cut the ligament and remove the right valve.

If you have not yet studied the inside of the shell go back to "Internal Shell Features" and do so using the right valve.

## MUSCLES

Look at the body of the animal in the left valve. Return the right mantle skirt to its original position and inspect its outer surface. This is the surface that would normally lie adjacent to the inner surface of the right valve. Several muscles originate on the valves and pass through the mantle skirt to their insertion. Some are adductor muscles inserting on the other valve whereas others are pedal muscles inserting on the foot. Find the large anterior and posterior adductor muscles. You have already seen their scars on the valves. These adduct the valves by bringing them together on the midline thereby closing the gape. No muscles antagonize the adductors, instead the elastic ligament opens the gape and stretches the adductors.

Unionoids also have a pair of pedal protractor muscles to assist in protruding the foot out of the gape and two pairs of pedal retractor muscles to withdraw the foot back into the mantle cavity before the adductor muscles close the gape. The protractors and retractors are antagonists and their origins and insertions must be arranged so they can have opposite actions. Both originate at scars on the inner surface of the valve and insert on the lateral surface of the foot and visceral mass. Protractor origins (on the shell) are ventral to those of retractors whereas protractor insertions (on soft tissue) are dorsal to those of the retractors.

Near the postero-ventral corner of the anterior adductor find the right pedal protractor muscle. It is a small, but conspicuous, cord of muscle fibres extending obliquely posteriorly and dorsally from its scar on the right valve. Its origin on the valve is ventral to that of the pedal protractor muscle. You can see the muscle by lifting the mantle skirt to reveal the foot. It is usually easy to see and is often pale, either tan or white.

The right anterior pedal retractor muscle is near the pedal protractor but is a little harder to find. Its distal end (origin) is contiguous with the posterior edge of the anterior adductor muscle. The posterior pedal retractor muscle can be seen immediately dorsal to, and contiguous with, the dorsal edge of the posterior adductor muscle. It extends to the dorsal edge of the foot. The mantle is attached to the inside of the valves by pallial muscles along the periphery of the mantle and by two small inconspicuous muscles extending from the dorsal mantle to the valve near the umbones.

## FOOT

The foot is a large, laterally compressed complex of muscles with a central hemocoel on the ventral midline of the visceral mass. It will be contracted in both preserved and fresh dissected specimens. In life it is capable of great extension, of which you see no hint in your preserved specimen. The ancestral bivalve foot was adapted for digging in soft sediments and that continues to be true of almost all North American freshwater bivalves, native and introduced, except *Dreissena*. The pedal retractor and protractor muscles originate on the shell and extend to their insertions on the foot.

## MANTLE SKIRT

The two mantle skirts are lateral outfoldings of the mantle, or dorsal body wall of the visceral mass. They extend ventrally on either side of the mantle cavity, visceral mass, and foot and underlie the valves. The right mantle skirt should presently be uppermost in your preparation and should be hiding most of the remaining soft anatomy from view. Move the right mantle skirt dorsally to reveal the mantle cavity, foot, and visceral mass. This will also uncover the left mantle skirt lying intact against the left valve. Avoid disturbing the fragile (in preservative) connection between the left skirt and left valve.

Use the dissecting microscope (8X) and good illumination to examine the ventral margin of the left skirt and note that it consists of four distinct longitudinal folds or ridges extending along its entire length. The parallel folds are separated by three parallel grooves. Use fine forceps and needles to demonstrate the presence of four side-by-side folds separated by three grooves. In most bivalves only three folds, inner, middle, and outer, are present but in unionoids one of the original three folds is divided into two folds to give a total of four. The thick, wide inner mantle fold can be recognized by the small papillae along its free edge. It is the muscular fold containing the radial pallial muscles that extend to the pallial line as well as longitudinal pallial muscles whose fibres parallel the mantle margin. It is not involved in secreting the shell. The groove separating this fold from the middle fold is shallow.

The next two folds are separated by the deep periostracal groove. The periostracum originates in this groove and in an intact specimen will still be connected here. New periostracum is secreted by the epithelium of the medial surface of the outer fold, deep in the groove. Look along the edge of the left mantle skirt for places where the skirt remains attached to the shell by a thin, transparent, yellowish-brown, cellophane-like periostracum. The periostracum emerges from the periostracal groove between the outer and middle folds. The attachment between periostracal groove and periostracum is very fragile, and easily destroyed in preserved specimens. It may not be intact in your specimen. The attachment is robust in fresh specimens.

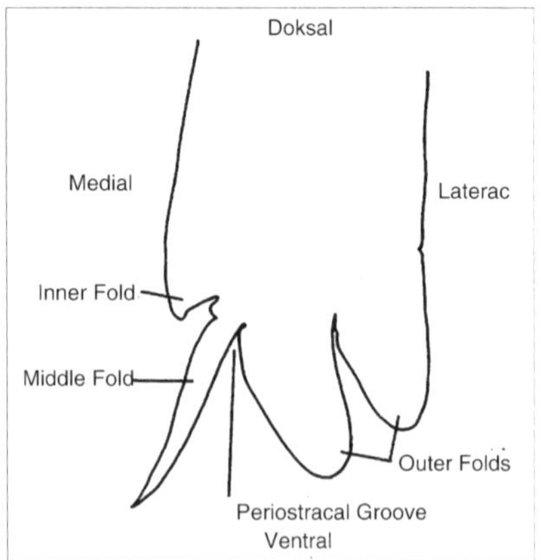


**Fig.** Cross Section of the Ventral Margin of the Mantle Skirt.

With forceps tug the periostracum out of the periostracal groove without damaging the edge of the skirt. Now, with magnification, examine the edge of the left skirt in a region from which the periostracum has been removed.

The thin middle mantle fold lies close to the outer mantle fold (once the periostracum has been removed) and the two may appear to be a single fold. Use a microneedle to tease the two folds apart and demonstrate the deep periostracal groove between the outer and middle folds. The periostracum is secreted by secretory cells of the outer fold at the bottom of this groove. The middle mantle fold is sensory.

The thick outer mantle fold (the one closest to the shell) is glandular and its outer epithelium secretes the prismatic layer of the shell. In unionoids the outer fold is actually composed of two folds. The entire outer surface of the mantle skirt secretes the lamellar (nacreous) layer of the shell. The mantle folds are not involved in secretion of the lamellar layer. At some point during the dissection the weakened (by preservative) attachments between muscles and shell in preserved specimens may fail so that the mussel falls out of the left valve. You may take advantage of this, if you wish, to view all aspects of the external anatomy without the interference of a valve. Be sure you are always able to replace the animal in its correct position and orientation in the left valve.

## VISCERAL MASS

The large central part of the body is the visceral mass. The foot occupies its median ventral border and the two mantle skirts arise from its dorsal margin. Most of the visceral organs, including the heart, kidney, gut and digestive ceca, and gonads, of a mollusc are contained within the visceral mass. The pigmented tissues of some these organs can be seen through the body wall at present but will be considered in more detail later.

The pericardial cavity occupies the extreme dorsal edge of the visceral mass between the posterior adductor muscle and the umbo. It is covered dorsally by a thin region of the body wall. This body wall is usually opaque in preserved specimens but may be translucent in fresh material. The pericardial cavity is the much reduced bivalve coelom, the chief body cavity of molluscs being the hemocoel.

The heart, consisting of two atria and a ventricle, is located in the pericardial cavity, as is a portion of the rectum. (If you are dissecting a living mussel the beating heart may be visible inside the thin body wall and pericardium.) Later in the dissection the pericardium will be opened to reveal its contents. The brown kidney lies close to the surface of the visceral mass lateral, anterior, and posterior to the pericardial cavity. The greenish digestive cecum (= digestive gland) shows through the body wall of the anterior visceral mass just posterior to the anterior adductor muscle. The yellowish gonad is situated in the central region of the mass dorsal to the foot.

## MANTLE CAVITY

### Chambers

The two mantle skirts enclose a large water space known as the mantle cavity. The gills form the roof of this chamber. In life position the mantle cavity is in communication with the external environment via two apertures, the inhalant and exhalant apertures.

On each side of the visceral mass and foot, a gill divides the mantle cavity into a ventral inhalant chamber (= branchial chamber) and a dorsal exhalant chamber (= suprabranchial chamber, epibranchial chamber, cloacal chamber, anal chamber). The gills are the floor of the exhalant chamber and roof of the inhalant chamber. The inhalant chamber is currently visible to you as the large space into which the gills protrude. Most of the exhalant chamber, however, is currently hidden. Looking into the exhalant aperture the space you see is the posterior end of the exhalant chamber. Water enters the inhalant aperture and flows into the inhalant chamber. It then passes through minute ostia in the gills to enter the exhalant chamber from which it exits through the exhalant aperture. The mantle cavity is a space outside the body of the mussel even though parts of it appear to be internal. It contains a current of river water that circulates through the mussel.

### Apertures

In unionoids the two mantle margins are largely independent of each other and are not fused except posteriorly where they form two apertures to channel water in and out of the mantle cavity. Together the right and left posterior mantle margins form the ventral inhalant aperture and the dorsal exhalant aperture. The right and left mantle skirts are fused together dorsal and ventral to the exhalant aperture. The exhalant aperture opens from the exhalant chamber of the mantle cavity to the outside. The inhalant aperture, although much larger than the exhalant, is not so obvious in a gaping specimen in which the mantle edges do not touch.

It is formed by thickened pads on the margins of the right and left mantle edges. When the valves are close together, so are these pads and they then form an opening, the inhalant aperture. Push the right and left posterior mantle edges together recreate the inhalant aperture. The dorsal margin of the inhalant aperture is formed by the fusion of right and left mantle skirts which also forms the ventral margin of the exhalant aperture. In contrast, there is no fusion of tissues to form the ventral margin of the inhalant aperture. The inhalant aperture is equipped with short chemosensory papillae but these will be contracted in preserved specimens. The exhalant aperture has no sensory papillae. The inhalant aperture opens from the outside into the inhalant chamber (= branchial chamber) of the mantle cavity.

By looking into the exhalant aperture you can see the posterior adductor muscle passing transversely across the exhalant chamber. The tubular rectum can be seen on the midline of the posterior adductor muscle. The rectum ends there at the anus, which you can also see by looking in the exhalant aperture.

Look dorsal to the posterior adductor at the junctions of the margins of the right and left mantle skirts. In most unionoids (*viz*, Ambleminae [sF] and Unioninae [sF] but not Margaritiferidae [F]) there are one or more supra-anal apertures between the two skirts. These openings are auxiliary exhalant apertures dorsal and anterior to the exhalant aperture. Use a blunt probe to demonstrate the continuity of the supra-anal aperture(s) with the exhalant chamber.

**Gills**

Observe that each gill (there is one on each side) is composed of two half-gills known as demibranchs. A whole gill is a holobranch. Each holobranch is derived from a single bipectinate gill of the ancestral bivalve and thus should still be thought of as a single gill. The mussel has two holobranchs (one right and one left) composed of two demibranchs each for a total of four demibranchs. Each holobranch consists of a lateral demibranch adjacent to the mantle skirt, and a medial demibranch adjacent to the visceral mass and foot. The holobranch is attached to the roof of the mantle cavity by a longitudinal central axis. The central axis lies between the two demibranchs.

Use a pair of scissors to remove the right mantle skirt with a longitudinal incision between the dorsal margin of the right lateral demibranch and the right mantle skirt. Start the incision immediately ventral to the posterior adductor muscle and avoid cutting into the region of the apertures. Extend the incision anteriorly and then dorsal to the labial palps and remove the skirt. This incision will open the exhalant chamber but the gill and labial palp will remain with the body. Modify the incision if necessary so that the exhalant chamber is open along the entire length of the gill. Do not open the pericardial cavity at this time.

Bivalve gills are composed of numerous slender gill filaments joined together to form sheets. Use 30X of the dissecting microscope to look at the surface of the lateral demibranch and you will see the very fine parallel gill filaments. Each filament begins at the central axis, drops down into the inhalant chamber then reverses direction sharply and climbs back up to the roof of the chamber where it attaches beside (lateral or medial) to the central axis. Each filament is attached to the filament anterior to it and the one posterior to it. Collectively all the filaments are joined together form a sheet, or lamella. Each demibranch is composed of two lamellae. One, the descending lamella is composed of the filaments that drop down from the central axis. The other, the ascending lamella, is composed of the same filaments on their way back up to the top of the inhalant chamber. Look at the lateral demibranch of the right gill. It should be facing you.

The surface you see is the ascending lamella of the lateral demibranch. Lift the demibranch and look at its other side. This is the descending lamella of the lateral demibranch. Look at its dorsal edge to see the central axis. With the lateral demibranch held up and out of the way you are looking at the descending lamella of the medial demibranch. It arises at the central axis. Finally, lift the medial demibranch and look at its medial surface. This heretofore hidden surface is the ascending lamella of the medial demibranch. Got it? Two holobranchs, two central axes, four demibranchs, eight lamellae in one mussel. Most bivalves, unionoids included, use their gills for suspension feeding as well as gas exchange. Both functions require that water (with dissolved oxygen and suspended food particles) enter the inhalant chamber, pass through tiny openings in the gill lamellae (between the filaments) into the exhalant chamber and then out the exhalant aperture. Oxygen and food particles are removed as the water passes through the lamellae. Adjacent filaments are held together by interfilamentary junctions but these are not unbroken continuous connections. Gaps in the junctions are known as ostia and provide a route for water to flow through the lamellae.

With fine scissors cut a 2 × 2 mm square of the ascending lamella and rinse both sides of it with a vigorous stream of water from a squeeze bottle. Make a wetmount and examine it with 400X of the compound microscope. Most of what you see will be filaments. Tissue interfilamentary junctions holding the filaments together will also be visible. The clear areas between the filaments, where the junctions are interrupted are ostia. Focus carefully on the edges of the filaments to see the lateral cilia extending from the filament into the ostia. These are the cilia that generate the feeding/respiratory current through the system. Other cilia, known as frontal cilia, are on the inhalant surfaces of the lamellae and are responsible for moving food particles and mucus over the gill. Frontal cilia may be difficult to see in this preparation.

Look again at the surface of the ascending lamella of the lateral right demibranch. Large evenly spaced ridges extend across the demibranch parallel to the filaments. These are tissue interlamellar junctions responsible for holding the ascending and descending lamellae of the demibranch together. The interlamellar junctions divide the exhalant chamber into vertical water tubes that arise in the demibranch and empty into the exhalant chamber. We have seen that the lamella appears finely corrugated because it is composed of filaments but coarser corrugations are also present and these are the water tubes. The demibranch was opened in the vicinity of the exhalant aperture earlier when you cut away the right mantle skirt. Find the openings to the water tubes inside the lateral demibranchTRTYY. Note that the tubes open into the exhalant chamber. The kidneys and gonad connect via through separate ducts with openings in the anterior end of each exhalant chamber. Both are small slits of which the nephridiopore is slightly lateral to the gonopore. Female unionoids retain their eggs and brood them in the water tubes. Note if eggs, embryos, or larvae are present and if so return later during your study of reproduction to make a wetmount of them.

## LABIAL PALPS

A large, leaflike labial palp is located on each side of the visceral mass and gill at the anterior end of the mantle cavity. Each palp resembles a small gill and like the gill is bilobed, being composed of two similar, demibranch-like half palps. The lateral half-palp is associated with the lateral demibranch of the whereas the medial half-palp corresponds with the medial demibranch.

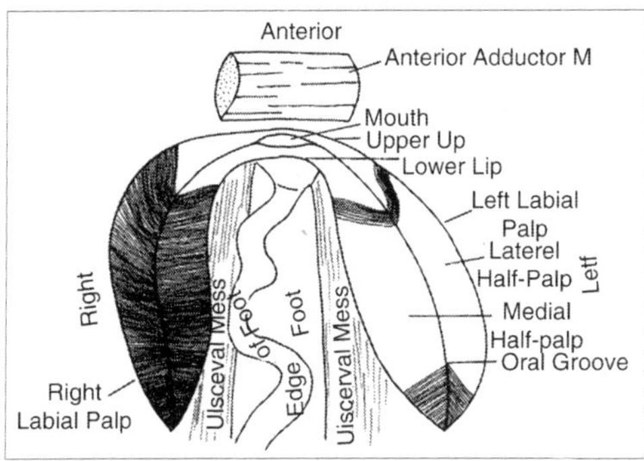


**Fig.** Ventral view of the Anterior end of the Foot and Visceral Mass. The Palps are Opened to Reveal the Sorting Fields. Most of the Ciliated Ridges of the Left Palp have been Omitted.

Find the right labial palp and lift the lateral half-palp to expose the medial half-palp. Particles and mucus are transported by ciliary currents from each demibranch to the associated half-palp for sorting. Examine the now-exposed surfaces of the half-palps with magnification to see that they are composed of an array of ciliated ridges and grooves.

These form a ciliary sorting field across which a string of mucus and food particles from the gills passes on its way to the mouth. While crossing the palpal sorting field mineral particles are imperfectly separated from organic food particles. The mineral particles drop off the edges of the palp into the inhalant chamber as pseudofeces whereas food particles continue on to the mouth.

In life the ciliated, ridged surfaces of the two halves of the palp face each other and the sorting field is not visible from the outside. In the crease where the two half-palps join is a ciliated oral groove that goes to the mouth. Mineral particles are transported laterally, away from this groove and food particles move along it. If the mussel is still in the left valve, remove it and set the valve aside. Hold the mussel so you can view the anterior end of the visceral mass and foot with magnification. Open the right palp so the sorting fields of both half-palps are completely exposed. Manipulate the mussel so you can trace the anterior ends of each half palp anteriorly towards the midline. The anterior end of the lateral half-palp tapers to a narrow ridge of tissue that extends anteriorly to cross the midline where it joins the similar anterior extension of the left lateral half-palp. The anterior extension of the right medial half-palp similarly joins with the left medial half-palp.

These two ridges cross the anterior midline of the visceral mass immediately dorsal to the foot and ventral to the anterior adductor muscle. Located between them is the large, open mouth. The ridge connecting the lateral half-palps is dorsal to the mouth and forms the upper lip. The lower lip connects the two medial half-palps below the mouth. The mouth is much easier to see in unionoids than in most other bivalves.

## INTERNAL ANATOMY

### Nervous System

The central nervous system is typical of bivalves and consists of paired ganglia connected by commissures and connectives. In life the ganglia contain neuroglobin which, if it has not faded in preservative, imparts a rusty orange, brownish, or yellow colour and makes them easier to locate.

The large visceral ganglion is easily found making it a good starting point for dissection of the CNS. The paired visceral ganglia are fused together on the midline to form what appears to be a single ganglion. It is on the ventral margin of the posterior adductor muscle and may or may not be visible through the thin epithelium covering the midline of the ventral surface of this muscle. Remove this epithelium to expose the ganglion. Three major pairs of nerves exit the ganglion. A pair of conspicuous cerebrovisceral connectives (= visceral nerves) connect it with the cerebral ganglia. The connectives can be traced anteriorly into the visceral mass for a short distance but anteriorly they are embedded in the mass and cannot be seen without careful dissection. A pair of large posterior pallial nerves from the visceral ganglia serve the posterior mantle margin. A pair of large branchial nerves from the visceral ganglion extend to the gills. The heart is served by a pair of small nerves from the visceral ganglion.

The smaller cerebral ganglia are situated anteriorly, one on each side of the mouth. They are more difficult to locate than the visceral ganglia. Find the mouth and the anterior pedal protractor muscle and pedal protractor muscle where they emerge from the visceral mass. The lips of the labial palps pass over this area which is separated from the lateral corners of the mouth by a few millimeters. The ganglia are not immediately adjacent to the mouth. Carefully remove the thin body wall from the area just described to reveal the ganglion. The pleural ganglia are fused with the cerebral ganglia and cannot be distinguished from them in gross anatomy.

The two cerebral ganglia are connected by the slender cerebral commissure which arches dorsally over the mouth. A cerebropedal connective extends from each half of the cerebral ganglion to the pedal ganglion. An anterior pallial nerve extends longitudinally along the anterior mantle margin after arising from the cerebral ganglia. Each half of the cerebral ganglion sends a nerve to the anterior adductor muscle.

The two pedal ganglia are fused on the midline in the dorsal edge of the foot, embedded in the foot muscles. They can be exposed only through a careful dissection beyond the scope of this exercise. The short pedal commissure is embedded between the two contiguous pedal ganglia and cannot be seen in gross view. Together these elements (cerebral ganglia, cerebral commissure, cerebropedal connectives, pedal ganglia, and pedal commissure) form a nerve ring around the anterior gut (esophagus). Motor nerves from the pedal ganglion the muscles of the foot. A pair of spherical statocysts is located lateral to each pedal ganglion.

The peripheral nervous system consists of sensory and motor nerves extending to and from the above ganglia. The most important have been mentioned.

**Hemal System**

The bivalve hemal system consists of a heart, arteries, veins, and a hemocoel consisting of large, blood-filled sinuses. The blood is colorless. The heart is located in the dorsal pericardial cavity mentioned earlier, and is easily demonstrated. The remainder of the hemal system is difficult to see in these specimens.

Relocate the thin dorsal body wall on the dorsal midline just anterior to the posterior adductor muscle. This part of the body wall covers the pericardial cavity. Use fine scissors to make a longitudinal incision through this thin body wall to open the pericardial cavity. The incision should extend from the posterior adductor muscle anteriorly to near the posterior edge of the umbo. Be careful that you cut only through the thin body wall and that you do not damage the organs within the pericardial cavity. The relatively large space this incision exposes is the pericardial cavity lined by a peritoneum known as the pericardium. It is a coelomic space and contains the heart and rectum. A large tube, the rectum, enters the cavity anteriorly, extends the longitudinally for length of the cavity, and then exits posteriorly to end at the anus on the posterior margin of the posterior adductor muscle. The rectum passes through the center of the heart ventricle.

The heart consists of a single medial ventricle and two lateral atria (= auricles), one on either side of the ventricle. The atria and ventricle are both contractile. The large and conspicuous ventricle is wrapped around the rectum and is easy to find, although its relationship with the rectum may surprise you. Blood exits the ventricle via an unpaired median anterior aorta lying on the dorsal surface of the rectum, and an unpaired median posterior aorta on the ventral surface of the rectum.

The aortae are difficult to demonstrate. The anterior aorta supplies the visceral mass, anterior adductor muscle, gonad, foot, anterior mantle, and kidney. The posterior aorta supplies the posterior adductor muscle, pericardium, rectum, and the posterior mantle. The venous return consists of veins and sinuses ultimately draining into the vena cava, a large sinus ventral to the pericardial cavity and between the two kidneys. Blood from the vena cava goes to the gills via the afferent branchial vessels where. Oxygenated blood leaves the gills via the efferent branchial vessels to the atria. The atria drain into the ventricle.

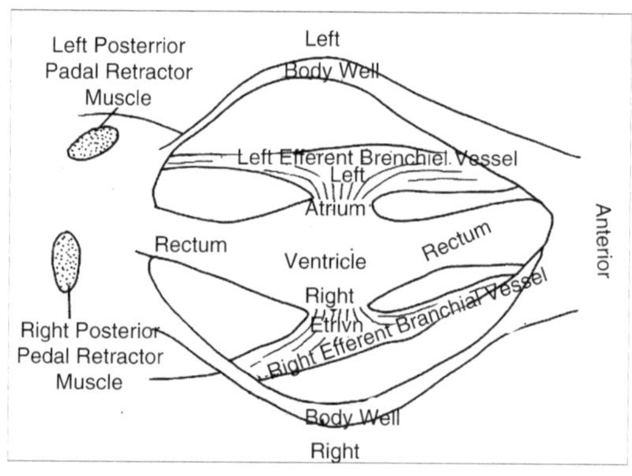


**Fig**. Dorsal View of the Opened Pericardial Cavity.

*Gently* pull the ventricle and rectum to the side, away from the body wall while watching under magnification. This will enhance your view of a thin, triangular, membranous atrium connecting the side of the ventricle with the membranous efferent branchial vessel from the right (or left) gill.

Use fine scissors to open the right side of the ventricle with a longitudinal incision along the entire ventricle. This will open the ventricular lumen for study. Observe the thick muscular walls of the ventricle.

Notice the conspicuous tubular rectum passing longitudinally through the lumen. Look on the right side of the ventricle for an atrioventricular aperture between the atria and ventricle. The aperture is guarded by an atrioventricular valve consisting of two thin, membranous, longitudinal folds of tissue arranged like lips above and below the aperture. Although large, these lips can be difficult to distinguish from the muscular ventricular walls and from each other. Use fine forceps and a microneedle to probe the tissues and demonstrate the lips. The two lips form a one-way valve that prevents the backflow of blood from the ventricle to the atria. You can see, by examining the configuration of the lips, how back pressure on the membranes would close the aperture.

Contractions of the atria propel blood through the atrioventricular aperture into the ventricle. Subsequent contraction of the ventricle closes the atrioventricular valve and forces blood into the anterior and posterior aortae.

## EXCRETORY SYSTEM

The bivalve excretory system consists of a pair or large, elaborate metanephridia, or kidneys, draining the pericardial cavity to the exhalant chamber. Remember that the pericardial cavity is a coelom. Each kidney is an elaborate tube extending from the nephrostome, opening from the pericardial cavity, to the nephridiopore, opening into the exhalant chamber. A special elaboration of the atrial wall, equipped with podocytes, is specialized for ultrafiltering the blood into the pericardial cavity to form the primary urine. The primary urine enters a nephrostome and passes through the kidney to the nephridiopore.

During passage through the kidney lumen the primary urine (ultrafiltrate) is modified, chiefly by reclamation of solutes, and the resulting final urine, mostly water, is released from the nephridiopore into the exhalant chamber. The role of the kidneys in freshwater bivalves, living as they do in a hyposmotic environment, is chiefly osmoregulatory. The kidneys constantly pump excess water out of the tissues and back into the environment. Ammonia, the chief end product of nitrogen metabolism, is lost by diffusion across the mantle and gill surfaces.

Each of the two kidneys (= metanephridium, organ of Bojanus, renal organ) is a large sac located ventral, lateral, and posterior to the pericardial cavity. With fine scissors make a longitudinal cut through the body wall along the dorsal margin of the right gill and ventral to the pericardium.

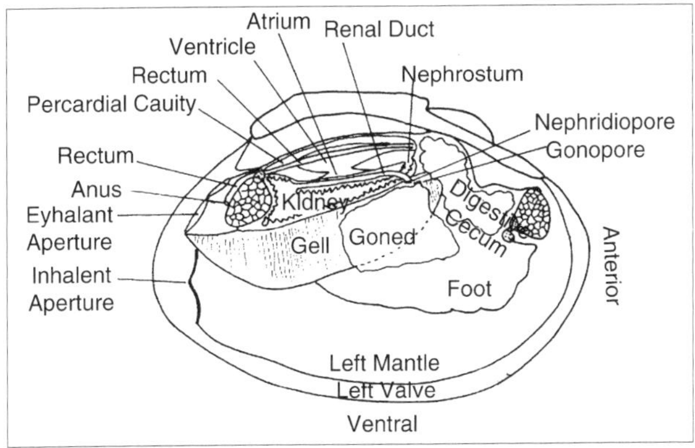


**Fig.** Dissected Unionoid Mussel Viewed from the Right side Showing the Pericardial Cavity and Kidney. The Gill and Visceral Mass are Drawn as if Transparent.

The kidney is recurved on itself and open at both ends. Proximally it is a large spacious glandular region that opens from the anterior pericardium via the nephrostome. The distal, downstream end of the kidney is a non-glandular renal duct that opens into the anterior exhalant chamber via the nephridiopore.

The ventral wall of the pericardial cavity is adjacent to the dorsal wall of the glandular portion of the kidney. If your incision was deep enough it will have opened the glandular kidney and revealed its lumen. The walls have a glandular appearance and consist of dark brownish gray tissue. The nephridiopore is a slit-like opening at about the level of the middle of the foot but it is not easy to demonstrate. The gonopore, either male or female, is nearby, slightly medial to the nephridiopore.

The two pericardial glands (= Keber's organs) are elaborations of the pericardial peritoneum equipped with abundant podocytes where ultrafiltration occurs. These brown organs can be seen through the body wall at the anterior end of the pericardial cavity. They receive blood from the efferent branchial vessel and ultrafilter it into the pericardial cavity to form the primary urine.

## REPRODUCTIVE SYSTEM

### Anatomy

Unionoids are gonochoristic with external cross fertilization. Some sexual dimorphism is present but it is usually not conspicuous (the shell of females of some species is more inflated than males to provide the space for brooding). The reproductive system consists of a single gonad, either ovary or testis, and two short gonoducts. The gonad, although a derivative of the coelom, is independent of the pericardial cavity.

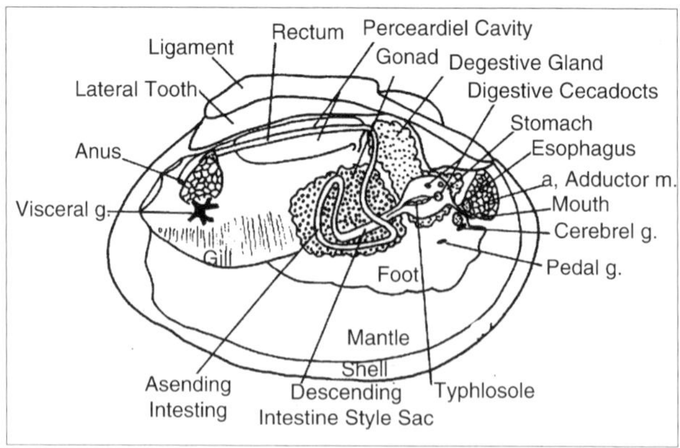


**Fig.** Dissected *Actinonaias* Showing the Gonad, Digestive System, and Ganglia. The star-shaped Visceral Ganglion is Shown a Little Larger than Life Size. A = Anterior, g = Ganglion, m = Muscle.

It does not share ducts with the kidney. It arises in bivalve evolution through the fusion of an ancestral pair of gonads. The two gonoducts, right and left, emerging from a single gonad, reflect this double origin. The gonad is large and occupies the space in the visceral mass extending ventrally from the kidney to the dorsal edge of the foot. Each gonoduct empties into the anterior exhalant chamber via a gonopore.

The shapeless gonad can be exposed by cutting away the thin body wall from the side of the ventral visceral mass at the top of the foot. It fills the space in the ventral visceral mass and is dorsal to the muscular foot.

## LIFE CYCLE

Females brood eggs and embryos to the glochidia larva stage in the water tubes. Females shed eggs into exhalant chamber from which they move into the water tubes of some or all demibranchs. In some (*viz. Amblema, Quadrula, Fusconaia, Gonidea, Tritogonia, Plectomerus, Quincuncina,* and *Megalonais*) both demibranchs on each side are used as brood chambers and are visibly swollen in brooding females but in the remaining Unionidae, including *Actinonaias*, *Lampsilis*, and *Anodonta*, only the lateral demibranchs are so used.

Sperm are shed into the river (or lake) from the exhalant aperture of a male individual. If caught in the inhalant flow of a female mussel, they pass through the ostia to enter the water tubes of the demibranchs. The waiting eggs are fertilized and retained in the water tubes where development begins and continues to the glochidia larva stage.

## GLOCHIDIUM LARVA

Females release glochidia larvae through the exhalant aperture. Glochidia are modified veliger larvae and are parasitic, requiring a fish host before metamorphosing into a juvenile mussel capable of independent existence. Glochidia parasitize and feed on a fish host before dropping off into the sediment to metamorphose into a juvenile mussel.

Inspect the water tubes for the presence of glochidia larvae. If present, make a wetmount with a wax supported coverslip and study it with the compound microscope. Glochidia are derived bivalve larvae with a pair of valves, adductor muscle, sensory bristles, and a larval thread. Species that parasitize the skin of their fish hosts also have a conspicuous hook at the ventral border of each valve. Species that attach to the gills lack the hooks.

## DIGESTIVE SYSTEM

The digestive system consists of mouth, esophagus, stomach with digestive ceca, intestine, and anus. Remove the mussel from both valves if you have not already done so. The large mouth is on the midline at the anterior end of the visceral mass flanked by a labial palp on each side. Hold the mussel with one hand and examine the anterior midline of the visceral mass to find the mouth. The mouth opens into the short esophagus which extends posteriorly to the stomach. Insert a blunt probe into the mouth and look into the esophagus while holding the mussel on the stage of the dissecting microscope.

With the mussel in a small dissecting pan of water on the stage of the dissecting microscope use fine forceps to dissect away the thin body wall posteriorly from the right side of the mouth. You may already have done some of this to expose the right cerebral ganglion. Removal of the body wall will expose the esophagus, which is a short but wide, very flat, thin-walled tube about the width of the mouth. With fine scissors open the esophagus posteriorly to the stomach. The esophagus is lined by a ciliated epithelium. Further dissection of the digestive system is destructive and should not be attempted until all other organ systems have been studied.

Extend the esophageal incision posteriorly to open the anterior stomach. Initially you need cut only the thin gut wall but to open the posterior stomach you must cut through a thick layer of the wall of the visceral mass. The stomach is surrounded by the two greenish digestive ceca which can usually be seen, without dissection, through the surface of the visceral mass posterior to the anterior adductor muscle. The ceca are diverticula of the stomach to which they remain connected by ducts opening from the stomach walls.

The stomach is a large chamber in the anterior dorsal visceral mass. With magnification examine the anterior folded walls of the stomach. Anteriorly are four openings to the digestive ceca. The largest opening is low on the left wall, another is to the right of it, and two others on the dorsal wall. All these openings are in the anterior half of the stomach. Insert one blade of your fine scissors one of the apertures and open the duct with which it connects. This duct will soon enter a greenish digestive cecum. The ceca walls are elaborated to form abundant acini, or pouches, lined with secretory and absorptive epithelium. Absorption and most digestion, both intra- and extracellular takes place in the ceca. Indigestible particles are returned to the stomach. Some extracellular digestion takes place in the stomach.

Much of the wall of the stomach bears the fine but conspicuous, parallel, ciliated ridges and grooves of sorting fields whose function is to separate incoming particles and send organic particles to the digestive ceca and mineral particles to the intestine.

On the ventral stomach wall is the large protuberant ciliated typhlosole whose role is shunt mineral particles wasted from the digestive ceca to the intestine. Note that a slender ridge continuous with the larger portion of the typhlosole exits the large ventral aperture.

Posteriorly the intestine and style sac share a common bilobed aperture from the stomach. This aperture is partially divided, by two ridges, into the intestine on one side and the style sac on the other. Notice that the typhlosole enters this opening. Most material entering the intestine is indigestible mineral particles and the chief function of the intestine is feces formation and storage. The ciliated, secretory epithelium lining the style sac secretes and rotates an elongate, flexible, pellucid rod, the crystalline style, which extends out of the style sac into the lumen of the stomach. The distal (stomach) end of the style rubs against a chitinous plate, the gastric shield, in the wall of the stomach. The style is composed of digestive enzymes secreted by the sac epithelium and is present only in individuals that are feeding or have recently fed. It is resorbed in starved individuals. Consequently it may not be present in your specimen. It is large an unmistakable when present.

Follow the lumen of the style sac and intestine into the gonad ventral to the stomach. The style sac and intestine extend, side by side with continuous lumina, into the visceral mass where they are surrounded by the gonad. Here the style sac eventually reaches a dead end and stops but the intestine continues on making several loops until it eventually reaches the anus. The intestine is divided into three regions but its path through the visceral mass is difficult to follow. Use your scissors to trace the intestine lumen as far as you can. Beginning in the posterior stomach insert one blade of the scissors into the opening of the intestine and style sac and cut completely through the thick wall of the visceral mass. Most of this wall is gonad. You will cut through digestive ceca (greenish) and gonad (yellowish) as you trace the gut.

The descending intestine exits the posterio-ventral end of the stomach and its lumen is beside and continuous with that of the style sac for the length of the sac. It and the sac are a straight bilobed tube extending into the gonad in the ventral visceral mass.

Deep within the visceral mass, the style sac ends and the intestine continues as the ascending intestine, or middle limb of the intestine. This second of three regions of the intestine loops through the gonad and then extends dorsally to exit the gonad. The walls of the middle intestine are thin, making this limb difficult to trace in its wanderings.

Having exited the gonad, the gut extends posteriorly as the rectum. The rectum, which you have already seen, enters the pericardial cavity, passes through the ventricle, exits the cavity, and curves dorsally over the posterior adductor muscle to end at the anus on the posterior side of the muscle. Insert a teasing needle into the slit-shaped anus to demonstrate its presence. The wall of the rectum is folded to form a large typhlosole that occupies most of the lumen. Relocate the rectum where it enters the anterior end of the pericardial cavity and trace it through the ventricle. With a longitudinal incision open the rectum and find the typhlosole.

# 6

# Causes of Deforestation and Forest Degradation

Forests are one of the most prominent geographic features of the world in which we live. They occupy nearly a third of the earth's total land area and occur wherever conditions are not too dry, too cold, or too barren to permit their growth. In addition to being one of the most conspicuous parts of the landscape, they influence the lives of human beings in many ways-mostly beneficial. They moderate local climates, reduce soil erosion, regulate stream flow, support a wide variety of industries and communities and afford unequaled opportunities for recreation. Some of these influences can be evaluated in dollars and cents; others cannot, but are none the less important. Let us look first at some of the influences in the latter class.

## Atmospheric Conditions

The climate of any given place is determined chiefly by its latitude, its elevation and meteorological forces that operate over wide areas and at high as well as low altitudes. But within the broad limits established by these major factors, local modifications of the "microclimate" are effected by topography and forest cover. Although these effects vary greatly with different combinations of slope, aspect, kind and size of trees, density of stand and area covered, certain generalizations as to their character are possible.

### WIND MOVEMENT

The first thing that one notices on entering a forest on a windy day is the relative calm. So great is the difference that one seems to have stepped suddenly into another world. The mechanical obstruction offered by the trees deflects upward a large part of the moving mass of air and slows down the velocity of that which enters the forest. Dense stands with heavy foliage naturally have the greatest effect. Thus, a forest of deciduous broadleaf trees has more effect on wind velocity in summer than in winter. Observations indicate that the velocity of the wind in the interior of a forest may range from about a tenth to two thirds of that in adjacent open country, depending on the character and condition of the forest. The stronger the wind, the greater the reduction of actual velocity, but not of relative velocity. Under normal climatic conditions wind velocities within a forest are typically low, seldom averaging more than a few miles per hour.

This influence on wind movement is felt not only inside the forest itself, but for a considerable distance to the leeward. Shelter belts consisting of several rows of trees are often planted to take advantages of this fact, particularly in regions where lack of natural forest cover gives the wind an unbroken sweep over long distances. Intensive studies in the plains region of the United States show that shelter belts may have a measurable influence on wind velocity to a distance of as much as fifty times the height of the trees. The greatest effect usually comes within a distance of from three to five times the height of the shelter belt, where the velocity of the wind in its lee may be 20 to 30 per cent of that in the open.

This reduction of wind movement brought about by forests and shelter belts adds to the comfort of man and beast and increases the production of crops grown to the leeward of their sheltering influence.

## TEMPERATURE AND RADIATION

Forests exercise a moderating influence on air temperature as well as on wind movement. Here again the extent of the influence is greatest in dense stands with heavy foliage, which have the maximum effect on incoming and outgoing radiation. Protection from solar radiation, which under extreme conditions may be in the forest only 1 per cent of what it is in the open, reduces maximum temperatures throughout the year. In the forests of the United States the reduction in monthly maxima may range from about 3° F. in January to 8° F. in July. Similarly the slowing down of outward radiation from the earth generally raises minimum temperatures throughout the year. The effect of a forest on monthly minima is less than on monthly maxima, both in summer and winter, but may amount to more than 6° F. in the United States.

Since maximum temperatures are normally lower and minimum temperatures higher in the forest, the net effect of a forest cover is to reduce materially the range in temperature at all times of the year and also to reduce slightly the mean temperature. Evergreen trees have more influence than those with deciduous foliage. Although the actual figures may seem small, it must be remembered that the effect on absolute maximum and minimum temperatures is much greater than the effect on mean temperatures; the moderating influence of the forest is most noticeable and most welcome on extremely hot or extremely cold days.

## HUMIDITY

In general, the air in a forest is somewhat moister than that in the open, although the difference in absolute moisture content is less than would be expected in view of the large amount of water transpired by trees. Relative humidity, which varies with temperature as well as with the absolute amount of water vapour in the air, may be as much as 11 per cent higher in the forest. As in the case of temperature, the influence of the forest is much more pronounced on maximum and minimum humidities than on mean humidities.

## PRECIPITATION

No subject in the field of forest influences has aroused more controversy than the effect of the forest on precipitation. Whether that effect is negligible or significant has often been argued with more heat than light. Popular belief in the efficacy of forests as "rainmakers" was formerly much more widespread than it is today. Passage of the Timber Culture Act of 1873 was undoubtedly facilitated by the hope that tree planting in the prairie region would increase the supply of water, as well as of wood. Unfortunately, so many complicating factors are involved as to make positive proof one way or the other extremely difficult.

The argument that forests increase rainfall and other forms of precipitation is based chiefly on the fact that they transpire enormous quantities of water. Trees not only use more water in the process of growth than does any other form of vegetation, but may even add more moisture to the atmosphere than is evaporated from an equal area of open water. In effect they act during the growing season as a pump, moving water from the soil to the air in almost unbelievable quantities. And whatever influence transpired water may have in increasing precipitation is strengthened by the lower temperature in and above forests, by the fact that they add to the effective height of the land and by the friction which they offer to wind movement.

All these factors, it is asserted, tend to result in more frequent and greater precipitation-an influence that may be regional as well as local. Raphael Zon, for example, in his report to the National Waterways Commission argued that "forests in broad continental valleys enrich with moisture the prevailing air currents that pass over them and thus enable larger quantities of moisture to penetrate into the interior of the continent. The destruction of such forests... affects the climate, not necessarily of the locality where the forests are destroyed, but of the drier regions into which the air currents flow."

The opposing argument is that the major forces controlling precipitation operate over such wide areas — continental or even intercontinental in extent — that the presence or absence of forests can have virtually no effect. Furthermore, precipitation depends not only on the amount of moisture in the air, but on conditions favorable to its release. How forests affect these conditions, if at all, is uncertain.

Many attempts have been made to measure the influence of forests on precipitation, with widely varying results and with even more diverse interpretations of their meaning. Although most of these studies show significantly greater precipitation in the forest than in the open, the difference may be more apparent than real. This is primarily because the greater wind velocity in the open reduces the amount of precipitation caught by rain gauges. There are also other complicating factors that make it difficult to obtain strictly comparable measurements.

The opinions expressed by Joseph Kittredge in his book *Forest Influences* probably represent the point of view held by most scientists today.

He concludes that forests have no appreciable effect on cyclonic precipitation, but that they may increase local rainfall in temperate climates by not more than 3 per cent and that rainfall in clearings in the forest may be about 1 per cent more than in similar situations in the open.

*Evaporation*: Forests tend to decrease the amount of water evaporated from the soil under their immediate cover and to some distance in their lee. This is due to their influence on wind movement, temperature and humidity. One of the major objectives in the planting of windbreaks or shelter belts is to reduce evaporation and thereby increase crop production in the fields that benefit from this protection, which, commonly extends over a distance many times the height of the windbreak.

## SOIL CONDITIONS

### TEMPERATURE

Soil temperatures are affected by a forest cover in the same way as air temperatures but to a much greater degree. Mean daily maximum temperatures at the soil surface may be reduced in summer by as much as 50° F. and absolute maxima even more. The influence of the forest varies inversely as the distance from the surface, but under some conditions may be recognizable to a depth of 30 feet.

Minimum temperatures show a similar but less pronounced increase. This influence is particularly noticeable in the winter, when soils usually freeze later and less deeply in the forest than in the open. Sometimes the soil in the forest remains unfrozen while that in adjacent open fields is frozen to a considerable depth. On the other hand, in situations where there is a heavy blanket of snow in the open and a very light blanket under a dense forest canopy, this relationship may be reversed.

These influences are due partly to the reduction of insolation and radiation by the overhead forest cover and partly to the insulating effect of the litter and humus of the forest floor. These reduce maximum temperatures more than they increase minimum temperatures and hence the net effect is a cooling one.

In both cases the effect is beneficial. During the summer young plants are protected from high temperatures which in exposed situations in the open may be so extreme as to kill the living tissue; while during the winter reduction in the depth and period of freezing diminishes surface run-off by permitting the penetration of more water into forest soils.

## COMPOSITION AND STRUCTURE

Forests have a marked influence on the composition and structure of the underlying soil. Every year they add to the forest floor large quantities of leaves, twigs and branches, which under normal conditions are constantly decomposing to form humus. Part of this humus gradually mixes with the mineral soil beneath and also supplies it with soluble compounds carried downward by percolating water.

The amount of organic material in the forest floor varies widely with differences in climate, soil and vegetation. Recorded data run from less than 2 tons per acre in old-growth longleaf pine in Florida to nearly 120 tons in a forest of birch, sugar maple and spruce in New Hampshire. In general, high temperatures result in its rapid decomposition and disappearance. Ordinarily there is a gradual transition from undecomposed litter through partially decomposed duff and humus to a mixture of organic and inorganic material and eventually to mineral soil. An exception occurs in coniferous forests in the north where slow decomposition of the needles results in acid "raw humus." This raw humus does not mix well with the underlying mineral soil and in extreme cases may be peeled off like a blanket.

The usual effect of a forest cover is to improve both the chemical and the physical characteristics of the soil. Nitrogen, calcium, phosphorus, potassium and other elements are made available in larger quantities. Even more important is the action of organic materials in making the soil more friable and crumbly. They tend to make light soils, such as sands, heavier; and heavy soils, such as clays, lighter. These effects are strengthened by the constant growth and death of the tree roots. While growing they keep pushing into and loosening up new areas of soil; when dead they add organic material to the soil and leave channels through which water may percolate more readily. Porosity of frozen soil is also increased in the forest, where it is more permeable than in the open.

## MOISTURE CONTENT

Forests tend to decrease the amount of moisture in the soil by intercepting precipitation, by retaining water in the forest floor and by transpiration. They tend to increase the amount by reducing surface run-off, by increasing the permeability of the soil and by decreasing evaporation from it.

Much precipitation in a forest never reaches the ground because it is first intercepted by and then evaporated from, the leaves, twigs, branches and trunks.

This loss may run as high as one tenth of an inch per shower. The amount of water that actually reaches the ground may vary from none in light showers to the great bulk of the total precipitation in heavy rains. Interception naturally increases with the density of the stand and of the foliage and is much greater with deciduous trees in summer than in winter.

When the precipitation reaches the ground it is first absorbed by the forest floor, which has a field moisture capacity (that is, an ability to retain water against the pull of gravity) of from one to five times its own dry weight and when saturated contains much larger quantities. Its absorptive and storage ability is much greater than that of mineral soil, the field capacity of which usually ranges from 10 to 50 per cent of its dry weight. The thicker and the more decomposed the litter and humus of the forest floor, the more water it will hold.

The amount of water that actually gets through to the mineral soil may vary from zero to 100 per cent, depending on the amount of precipitation and on the condition of the forest floor. With light showers in times of drought all of the precipitation may be retained by the forest floor; while with heavy showers and in wet periods little or none may be so retained.

On land without vegetative cover and particularly with heavy soils, the surface layer tends to become hard and impermeable, with marked reduction in the amount of precipitation that sinks into the main body of the soil. In the forest, on the other hand, the litter slows down surface run-off and thus gives more time for the water to be absorbed by the humus, which acts as a huge sponge. More important still, the humus prevents muddying of the underlying mineral soil, which remains porous and thus absorbs and retains much more water than does exposed soil in the open.

Enormous quantities of water are removed from the soil by transpiration. A well-stocked forest of mature trees uses more water than does any other form of vegetation. It may require from 100 to 1,400 pounds of water for every pound of dry matter produced. Transpiration is far greater in summer than in winter and during the growing season is somewhat greater with broadleaf deciduous trees than with conifers. Its tendency is obviously to reduce the moisture content of soil in the forest as compared to that in the open. On the other hand, the reduction of evaporation in the forest, which was mentioned in the discussion of atmospheric conditions, tends to maintain the moisture content at a higher level than in the open.

The net effect of these conflicting influences varies widely with the character of the topography, the soil, the climate and the forest. So far as any generalization is possible, it may be concluded that soils under a forest cover tend to be somewhat drier than similar exposed soils in the open on fairly level ground and somewhat moister on steep slopes where more water is lost by surface run-off in the open than by interception and transpiration in the forest.

## Erosion and Stream Flow

Of all the influences of the forest on its environment, the most obvious and the least controversial is the reduction of surface run-off of water. This prevents or substantially checks soil erosion and modifies stream flow.

### ACCELERATED EROSION

Soil is constantly being moved by water and gravity from higher to lower elevations. Under normal conditions, with an undisturbed vegetative cover, this is a slow process known as "geologic" erosion. Over the centuries it has built up fertile alluvial soils in the valley bottoms. With the removal or disturbance of the vegetative cover, a striking change takes place. Larger and larger particles of soil and even huge boulders, can now be moved and the rate of removal is greatly increased.

The beneficial process of geologic erosion is replaced by the destructive process of "accelerated" erosion. Among the harmful results are deterioration or ruin of the lands where the erosion takes place and where the coarse detritus is deposited, siltation of reservoirs, impairment of the quality of water for municipal and industrial uses, destruction of fish habitats, clogging of river channels and increase in the volume of floods.

All kinds of vegetation serve to check erosion, but by far the most effective are well-stocked forests and well-sodded grasslands. Their influence is primarily due to their ability to reduce the amount and velocity of surface run-off.

The amount of material that water can carry at a given velocity is directly proportional to its volume. As the velocity increases, however, there is a much more than proportional increase in the cutting and carrying power of a stream. Thus, if the velocity of water is doubled, its cutting power is increased fourfold, its carrying power thirty-two-fold and the size of the material it can carry sixty-four-fold. These facts explain the occurrence of "mud-rock" flows, the transportation by small streams of boulders weighing many tons and up to a six-thousand-fold increase in the rate of erosion following destruction of the forest and the forest floor.

During ordinary storms much, if not all, of the precipitation is absorbed by the forest litter and humus, retained a short time and then passed on gradually to the mineral soil beneath. The humus layer is not comparable, as has sometimes been alleged, to a blotter on a slate roof, the efficacy of which in storing water ceases as soon as it becomes saturated. Except in "raw humus," there is no sharp line between the organic surface and the inorganic subsurface material in the forest.

On the contrary, there is a gradual transition from the top layer of newly fallen, undecomposed litter to partially and completely decomposed humus, to a decreasing mixture of organic with inorganic material and finally to mineral soil. Thus there is a constant and uninterrupted downward movement of water from the temporary storage provided by the upper layers of the forest floor into the larger and more permanent soil reservoir beneath. From there the water moves slowly as a subsurface flow to springs, streams and lakes, from which it is in time evaporated and again precipitated in the course of the hydrologic cycle.

Tree roots also help to keep the soil from being washed away in gullies and along the banks of streams and tend to check the occurrence of landslides. Erosion by wind is virtually non-existent in the interior of a forest and is greatly reduced for some distance to its lee.

In many ways forests are influential in preventing the development of accelerated erosion — the greatest threat to the world's soil resources. Fortunately their influence is most effective where the danger is most acute, as on steep slopes, in clayey soils and with heavy rainfall. Wind erosion can also be prevented or greatly reduced by properly placed windbreaks in open country, as in the Great Plains of the United States.

## DISTRIBUTION OF RUN-OFF

For all practical purposes the total stream flow from any watershed consists of the precipitation less the sum of the losses from interception, transpiration and evaporation. To the extent that the forest increases interception and transpiration, it tends to reduce stream flow; and to the extent that it decreases evaporation, it tends to increase stream flow. The net effect depends on the relative weight of these influences. It will obviously not be the same under all environments or at all seasons of the year.

Measurements from forested and from deforested or partially forested areas in Switzerland, Colorado and North Carolina have shown a strong tendency towards smaller annual run-off from the forested areas. The difference was particularly marked in the spring, when melting snow and heavy rains caused the maximum seasonal run-off. The greater discharge of water from the non-forested areas also carried with it much more eroded material.

The total flow of a stream is made up of two main parts: the surface run-off and the subsurface or ground-water run-off. The forest has a profound effect in reducing surface run-off and thereby increasing the amount of water available for subsurface run-off. The evidence is conclusive that in most regions surface run-off is very small or negligible from areas of undisturbed vegetation, while it may amount to half of the precipitation where the vegetation has been destroyed or seriously disturbed. However, the outstanding influence of forests in this respect is greatly weakened when the forest floor is destroyed by repeated fires, even though the trees themselves are not killed.

Reduction of surface run-off results in more uniform flow of a stream throughout the year. Maximum flow is nearly always less and minimum flow usually more from forested than from comparable non-forested areas. The tendency is to avoid sharp peaks of flow in the spring and deep troughs in the summer. The total amount of flow may be less, but its greater uniformity is highly desirable.

## FLOODS

What effect these influences of the forest have on floods is a hotly debated question. Certainly forests cannot prevent an abnormally high flow of streams when heavy precipitation falls on saturated or frozen soils. If the storage basin is already full, there is nothing for any additional supply of water to do but to flow off by way of the nearest stream. On the other hand, when the soil reservoir is well below the saturation point, there may be far less surface run-off from forested areas than from those not covered by vegetation, even in severe storms.

By and large there can be no doubt that forests tend to reduce flood flow in small watersheds in hilly country. One would expect them to have a similar influence on large rivers made up of the flow of many small streams. Here, however, the situation is complicated by many factors, the combined effect of which it is difficult to evaluate. For example, the way in which precipitation is distributed over a river basin as a whole and the rate of flow of the tributary streams largely determine whether their peak discharges reach the main river simultaneously or in successive waves.

Obviously the first combination will result in a much higher flood than the second, even with the same or perhaps even a smaller total run-off. Therefore, the clear-cut influence of the forest on the flow of streams from small watersheds at the headwaters of large rivers may not always be reflected in the flow of the main river.

Nevertheless, the general tendency will be for the forest to decrease both maximum and total run-off and to increase minimum run-off, throughout the river basin. The action of the forest in reducing erosion also has an important influence on flood flow and flood damage.

Small streams with steep slopes issuing from deforested or burned-over watersheds may carry several times as much solid material as water. One instance is on record in southern California where solids comprised 88 per cent of the total flow. As the stream flattens out, the larger and heavier materials are, of course, deposited; but even rivers with small gradients may carry heavy loads of sediment. This material adds significantly to the volume of the stream and increases its erosive power, particularly in time of flood. It may, in fact, be even more important than the volume of water as a source of damage.

## WATERSHED MANAGEMENT

From the point of view of water supply, the ideal management of the forest, as a part of the broader field of watershed management, is that which will produce the maximum total run-off, well distributed throughout the year, with minimum erosion. Such management may not always be identical with that which will produce the largest supply of wood. Serious conflicts will, however, be the exception rather than the rule. When they do occur, the method of management to be selected must be based on relative values, among which water for domestic and industrial use, irrigation, power, navigation and recreation may rank high. Striking evidence of the basic importance of conserving both soil and water resources is afforded by the greatly increased attention being paid by such agencies as the Soil Conservation Service, the Tennessee Valley Authority and the Forest Service in the United States to the promotion of more effective watershed management, in which forest planting and improved forest practices play a prominent part.

## Wood, the Universal Raw Material

No less prominent than the influence of the forest on climate and run-off is its influence on industry. In fact, its tangible products affect our economic activities even more clearly and more directly than do the intangible services that it renders. They provide opportunity for the profitable employment of labour and capital in many important industries and furnish a wide variety of indispensable consumers' goods.

Wood is sometimes known as "the universal raw material" because it is so widely distributed and can be used for so many purposes. Although for some uses it has been largely displaced by other materials, there are few articles into the composition, manufacture, or transportation of which it does not enter. While perhaps no more "indispensable" in modern civilization than many other materials, such as iron, cement and glass, its range of use is certainly much broader than most of them.

For thousands of years, until the advent of coal, oil and gas, wood constituted man's chief source of fuel. Even today more than half of the total consumption of wood in the world goes for this purpose. Another third in the form of lumber and structural timbers is used for the construction of buildings of all kinds. Except in urban centers and where forests are scarce or lacking, most people continue to live in wooden houses.

Lumber and dimension stock are in turn remanufactured into innumerable articles such as toothpicks, toys, sporting goods, musical instruments, handles, boats, caskets and furniture. In the United States approximately an equal amount of lumber (about one seventh of the total cut) goes into boxes, crates and dunnage used in the shipment of commodities. Large additional amounts of wood are used for railroad ties, poles, piling, fence posts, mine timbers, excelsior, shingles, cooperage, veneer and plywood. Modern adhesives are greatly helping to expand the use of wood in the fields of ply and laminated construction and are making practicable the salvage of much low-grade material. Advances in engineering techniques now make possible the use of wood in members requiring high strength and long spans, where metal was formerly regarded as indispensable.

Even more spectacular advances have taken place in the chemical utilization of wood. The demand for paper and paper products seems unlimited. Each year larger and larger quantities of wood go into the manufacture of newspaper, book and magazine paper, writing paper, wrapping paper, wallpaper, building paper, pasteboard cartons and paper board. From "dissolving pulp" come cellophane, rayon, artificial wool and many plastics. Today one could, if one chose, be clothed wholly in textiles that originated in the forest.

A lively imagination is required to visualize the many other chemical substances that can be derived from wood. Among these are wood gas, potash, charcoal, acetone, methyl alcohol, sugar, ethyl alcohol, yeast, salicylic acid and vanillin. Potentially the forest is the source of large quantities of motor fuel and food.

During World War II, when gasoline was scarce, many automobiles in Europe were operated entirely on wood gas, which is still used to a considerable extent in trucking. Another potential substitute for gasoline is ethyl (grain) alcohol produced by the fermentation of sugar, which in turn has been produced by the hydrolysis of wood cellulose. The alcohol can also be converted into lubricating oil and synthetic rubber.

Wood sugars have already been used extensively in Europe as feed for livestock and are a possible source of food for human beings. They can also be turned into yeast of high nutritive value by inoculating them with the appropriate organisms.

Edible carbohydrates and proteins are among the innumerable products obtainable from wood cellulose, which in its original state is a highly indigestible product. Lignin, the other main constituent of wood, is a comparatively unknown substance, of which little use is now made but which is likely in time to provide a wide variety of important chemical products.

Many substances other than wood and its derivatives come from trees and are ready for use with relatively little further processing. Among these are nuts, maple syrup, chewing gum, tannin, dyes, rubber and resin. The last yields turpentine, universally used as a solvent in paints and varnishes and in the production of synthetic camphor; rosin, used by violin players and baseball pitchers and in the manufacture of soap, varnish and paper; pine oil, used in the flotation process of separating metals from their ores and in the textile industry in the fixation of dyes; and pitch, for which chemists are just beginning to find profitable uses.

Many pages would be required merely to list the useful substances that come from trees. Wood, once regarded by some as obsolescent, is now used more widely and for more purposes than ever before. Its inherently desirable properties of high strength in relation to weight, its workability, elasticity, non-conductivity of heat and electricity and beauty can be enhanced and its undesirable properties of shrinking and swelling, inflammability and susceptibility to attack by insects and fungi can be mitigated by methods developed by modern technology. Wood as wood, properly treated and handled, is one of the most useful materials known to man; while cellulose, lignin and resin are a veritable treasure house of chemical products already in use and yet to be discovered.

The significance of forest products is emphasized further by the facts that forests are both renewable and more efficient as producers of solid substance (ligno-cellulose) than is any other crop. Here is at least one case where we can have our cake and a rich cake at that and eat it too. Industrially the age of the forest is ahead of us as well as behind us.

### Forest and Wood-Using Industries

This situation makes it possible for man, through intelligent handling of the forest resources, to raise his standard of living by the continued and increasing production of a host of valuable, often indispensable, raw materials and finished commodities. It also provides widespread opportunities for the profitable employment of land, labour and capital.

## FOREST INDUSTRIES

Forests also constitute one of the major landscape features of the modern world. A much larger percentage of the total land area is primarily suitable for the production of forested crops than of harvested agricultural crops. This area, averaging about 30 per cent for the world as a whole and running up to 80 per cent or more in some countries and regions, supports a large number and wide variety of forest and wood-using industries. First comes the task of growing the trees — the practice of forestry. This involves the protection of the forest from fire, insects, disease and trespass; the determination of present stands and future yields; the conduct of cultural operations, such as thinnings, to speed up the growth and improve the quality of the stand; and the final harvesting of the mature crop so as to replace it by a well-stocked new crop of desirable species, including where necessary the production and planting of nursery material.

Cultural and harvesting activities necessitate logging operations, with the transportation of the resulting products from the stump to the mill, the railroad, or the point of consumption. Housing for woods workers must be provided and primary and secondary woods roads must be constructed. Next, for forest products not to be used in their original form as fence posts, mine props, telephone poles, piling, etc., come primary and secondary manufacturing. Saw mills, veneer mills, planing mills, box mills, pulp and paper mills, furniture factories, wood-turning factories and a host of other plants play their part in the preparation of both rough and finished goods for use by the ultimate consumer. All of these operations require large numbers of workers and large amounts of capital. Wherever forest supplies are abundant, the forest and wood-using industries rank high in both respects in national economies. For example, in the United States the chief manufacturing industries using wood (excluding the extractive industry of logging) account for about 10 per cent of the workers, the wages paid and the value added by manufacture in all manufacturing industries. Capital investments are particularly heavy in the pulp and paper industry, where costly machinery is necessary.

One of the most striking features of the forest and wood-using industries is their widespread geographic distribution. Logging must obviously be conducted where the forests occur. Since logs are such a bulky commodity, it usually pays to cut transportation costs by subjecting them to at least primary manufacture near the point of origin. As a result, wood-manufacturing plants are widely scattered throughout the forest regions of the world. In the United States they comprise about a fourth of the total number of manufacturing establishments and rank next in number to those concerned with food, apparel and printing and publishing.

## FORESTS AND COMMUNITY DEVELOPMENT

Because of their large number and wide distribution, the forest and wood-using industries influence the economic and social life of a great many communities. This influence is most keenly felt in the smaller cities and towns, particularly in wellforested regions, where they often constitute the dominant interest. In such situations, the prosperity of the community rises and falls to a very considerable extent with the prosperity of the forest-based industries. These have a decidedly stabilizing influence wherever the forests are managed on a sustained-yield basis, with the average annual cut and average annual growth in balance. Ghost towns do not occur where sustained yields are obtained and where mill capacity is adjusted to the growing capacity of the forest.

One of the important aspects of any industry is its influence on other industries and on service activities. The forest and wood-using industries purchase large quantities of axes, saws, tractors, trucks, planers, lathes, digesters, paper machines and other items of equipment; and they sell lumber, veneer, plywood, boxes, pulp, paper and similar items to other industries for use in the manufacture and marketing of their products. Railroads in some regions depend for their very existence on the freight revenue derived first from bringing raw materials, equipment and general supplies to communities dominated by one or more wood-using industries and secondly from transporting the products of these industries from mill to market.

Industry in general also supports many services, such as those supplied by grocers, druggists, clothiers, doctors, lawyers, ministers and teachers. The wood-using industries do not differ in this respect from other industries, except that in many rural districts they provide the sole or the main reason for the existence of a community. Where this is the case, they create the demand for such services; and where other industries are also present, they intensify that demand.

One further item of community concern is the relation of the forest and its dependent industries to the small landowners and especially to the farmer. The existence of a nearby wood-using industry provides a market, which in turn should encourage thrifty management of the wood lot. The additional income thus available can have a stabilizing effect both on the farmer and on the community of which he is a part. Forestry is a key to permanence.

## FORAGE AND WILDLIFE

Not all forest industries are based on the utilization of trees. In many regions forage produced in openings in the forest and sometimes under the trees themselves, provides feed for cattle, sheep and pigs supplementary to that which they obtain from pastures and the open range. Beef, mutton and pork thus become in part products of the forest, the use of which for grazing may furnish an important means of support for the livestock industry. Timber production and livestock production are, however, not always compatible and stock must be kept out of areas where grazing is injurious to the forest. Browsing and trampling are likely to be particularly harmful in hardwood forests, from which domestic livestock should ordinarily be excluded.

Many fur-bearing animals which spend at least a part of their lives in the forest provide the raw material for another important industry. Milady must have furs; and mink, otter, fisher, marten, weasel, beaver, fox, muskrat, rabbit and many others supply them for her. Their production can well be one of the objectives of multipurpose forest management.

The taking of other forms of wildlife by sportsmen and fishermen helps to provide a market for manufacturers of arms, ammunition and fishing gear. Users of the forest for recreational purposes step up business for makers of automobiles, cameras, clothing, insect-repellents and gadgets of all sorts; for keepers of service stations, hotels, motels and dude ranches; and for packers, guides and wranglers. This stimulation of economic activity may extend as far back as the iron mine, the petroleum well and the cotton field.

### Recreation, Inspiration

The most important influence of the forest as a source of recreation is not indirectly through industry, but on the recreationist himself. Its chief significance lies in the enhancement of human values rather than monetary values. The forest provides not only recreation, but "re-creation." In the hurly-burly of modern life, with its economic pressures and nervous tensions, people need occasionally to get back to nature. Where better than in the open and particularly in the forest, can they regain both physical and spiritual strength? As civilization has advanced, man's relation to the forest has continually changed. To primitive man the primeval forest was generally a gloomy and forbidding place. He viewed it with mixed feelings of fear and superstition. Tree worship was common in many tribes. "The groves were God's first temples"; they were also the first temples of pagan gods.

Gradually, as tools were developed and new uses for wood were found and as the demand for land for agriculture increased, man pushed back the forest frontier. Domination of the forest by man was a slow and difficult process which developed qualities of courage, hardihood, resourcefulness and ruthlessness. The men who lived and worked in the forest and whose hero was Paul Bunyan, were no weaklings.

Today throughout much of the world the early relationship is reversed. Virgin forests, except in parts of South America, Africa and Asia, are a rarity. There is a shortage rather than a surplus of wood for industrial use. A rapidly increasing population is becoming more and more concentrated in urban centers. Even the lumberjack usually lives in town or in a comfortable camp quite unlike the crude shacks of earlier days. Forest areas of outstanding value for recreation are shrinking even as the need for recreation becomes more acute.

Different people seek re-creation of body, mind and soul in different ways. Some like to hunt and fish with gun and rod; others prefer to use field glasses and camera. It is now the wild beast that fears man, not man the wild beast. Some like to lose themselves on camping trips in an inaccessible wilderness; others are satisfied to enjoy the beauty of the forest from the seat of an automobile. Many wish enough "roughing it" to give them the illusion that they are emulating the prowess of their forebears. Some seek health, others inspiration. All are attracted to the forest by the opportunity to find rest or strife, relaxation or stimulation, in an environment that takes them out of the stifling routine of their daily lives.

From time immemorial the forest has influenced the work of architects, painters, poets, dramatists and preachers. It has molded the character not only of those who live and labour in it but of those who visit it only occasionally to play or to pray. In the realm of the human spirit its gifts are invaluable both figuratively and literally.As Robert Marshall expressed it: "The most important values of forest recreation are not susceptible of measurement in money terms.

They are concerned with such intangible considerations as inspiration, esthetic enjoyment and a gain in understanding. It is no more valid to rate them in terms of dollars and cents than it would be to rate the worth of a telephone pole in terms of the inspiration it gives." In face of mounting pressure for more wood for homes, furniture, magazines, rayon clothing and other tangible goods, it is important in the management of the forest to remember that "man doth not live by bread only."Forests are an inescapable geographic fact. Not only do they dominate the landscape throughout much of the world, but they influence man's life in countless and often unsuspected ways.

They temper the climate; improve the soil; reduce erosion and floods; support a wide variety of important industries; provide opportunities for the profitable employment of land, labour and capital; stabilize communities; and offer a unique source of recreation and inspiration in close communion with nature. Their management in ways that will assure the perpetuation of these values is an urgent task, of supreme importance to the people of the globe.

## MEDICINAL PLANTS AND FORESTRY: PRINCIPLES AND PRACTICES

Over much of the earth's land surface forests represent the climax form of vegetation and, if left undisturbed, are the dominant feature of the landscape. These natural forests are generally looked upon as one of the great resources of the world — the bounty of nature, which down through the ages has contributed much to man's comfort and enjoyment as well as to his economic progress. They have not only provided wood and other products to meet his domestic and industrial requirements, but have often had an important and usually beneficial influence on his environment, particularly in the regulation of stream flow and the maintenance of stable soil conditions.

A feature of fundamental importance, which distinguishes forests from some other resources, is that they are renewable. This means that the mature forest, as it is cut, may be replaced by a new crop of trees; thus a continuous yield of wood is provided from the same area of land. The application of this principle is the primary basis of forestry practice, which may be defined briefly as the growing and using of successive crops of timber.

As the practice of forestry develops, the forests often come to be looked upon less as a natural resource and more as a crop that is dependent in large measure on man's knowledge and skill. However, forestry is not concerned exclusively with growing timber as a crop; forests contribute in many other ways to the well-being of mankind and proper forestry practices assure the maintenance and enlargement of these benefits.

Man is not a conservationist by nature: history shows that as long as the existing forest appears to be adequate to the demand, it is subject to exploitation for immediate revenue and there is little concern for its future productivity. In other words, forest conservation, involving some investment of time and money, only begins when it is apparent that current practices are endangering the resource with critical effects upon the supply of forest products and services.

Down through the centuries, as man has increased in numbers and spread over the face of the earth, he has caused and is still causing widespread destruction of the forest cover, partly of necessity to provide land for agriculture, settlement and industrial development, partly through ignorance and greed and partly in the belief that the forest resource is inexhaustible. This latter viewpoint is to some extent justified in the early development of countries and regions that have extensive forest wealth and relatively small populations. Here the forest is often looked upon as a hindrance to settlement and agricultural development, with the result that areas cleared may exceed requirements and include lands quite unsuitable for the purpose in mind.

Man has in fact retained under intensive agriculture only a small portion of the area that was originally cleared for this purpose. It is estimated that, as a result of his activities, man has destroyed about two thirds of the world's original forest cover; the remaining forests containing merchantable timber today occupy less than one fifth of the world's land area.

In some parts of the world forest destruction has been unchecked, despite the adverse effect that this has had on the general economy and on the living standards of the people. Asia, in particular, has suffered from the effects of forest clearance and forest destruction, with widespread soil deterioration, inadequate supplies of wood to meet basic domestic needs and low living standards for a large portion of the population.

Elsewhere, as the danger became apparent, methods of managing the forests to ensure their continued productiveness were evolved and thus the science of forestry came into being. However, the philosophy of conservation has developed slowly and forestry has invariably been the child of necessity: only after periods of forest destruction and land clearance has the need for husbandry and planned forest economy been realized.

The transition from exploitation to management takes time and requires the solution of many problems, technical, economic and administrative. In the beginning it may be characterized by a strong reaction against spoliation aimed at preserving the forest intact — "locking it up" — for posterity. In general, however, this is the antithesis of sound management of a renewable resource, which for its fullest development requires periodic harvesting and regeneration; an exception would be the preservation of representative samples of natural forest for scientific study and as "wilderness" areas.

Modern forestry originated in Germany and France and much of its development during the past two centuries has taken place in northwestern Europe. Attention to forest conservation in these countries came early owing to an increasing demand for timber and firewood and the difficulty of securing wood cheaply from other countries. These conditions and the serious effects of deforestation on stream flow, particularly in montane regions, made evident an urgent need for forest management. During the nineteenth century expanding populations and the iron and coal economy of the industrial revolution created new demands for wood and gave added impetus to the development of forestry.

Slowly and usually only as the need became urgent, other countries have followed the lead of western Europe and developed forestry practices to meet their own requirements. Such developments usually begin with restriction on the use of the forest, as in the setting aside of timber reserves and with protection of the forest against its more serious enemies, particularly fire. These steps are followed by efforts to secure new growth after cutting, by encouraging natural reproduction and by planting: with this the practice of silviculture begins. Finally, the accessible, productive forests are managed under cutting regulations and suitable silvicultural techniques to ensure a continuous yield of timber in perpetuity. Such forests are said to be managed on a "sustained-yield" basis.

While the natural forest usually provides the starting point for a forestry programme, it may not represent the optimum conditions that an area is capable of and its perpetuation is not always looked upon as the ideal objective of forest management. The aim, rather, may be to improve the natural forest as far as possible within the limits set by the local environment, silvicultural knowledge and economic considerations and in view of the purpose for which the forest is to be used.

Attempts to go beyond these limits and completely change the natural forest pattern have sometimes failed. A well-known instance was the century-long attempt to convert the lowland hardwood forests of Saxony to pure Norway spruce.

Despite early rapid growth, the spruce was often badly damaged by snow and frost as well as by fungal and insect attacks. By the end of the third crop of spruce, conditions had so deteriorated that hardwood species had to be reintroduced and it was generally conceded that the plan had been unsuccessful. On the other hand, there are many examples of the successful introduction of types and species of trees different from those in the natural forest. The success of these operations depends largely on careful study of the ecological factors involved.

The aim of forestry may vary from one country to another and even between regions within the same country, depending upon local conditions. Usually the primary purpose is to grow sufficient wood to meet a country's requirements — first for domestic consumption, then for export to foreign markets. Products other than wood, usually referred to as minor forest products, may sometimes be an objective of management. These vary in importance and kind from one country to another, *e.g.*, turpentine, resin, oils, fruits, maple syrup, bark for tanning, etc. The southeastern United States and the Landes in Gascony, France, are two regions where the production of resin and turpentine from species of pine is a major industry carried on in association with timber production.

While the production of wood and other commodities is often primary, in its broader aspects forestry is concerned with the interaction between the forest and its environment, as, for example, in the effects of the forest on the stabilization of soil, on the maintenance and restoration of soil fertility, on the regulation of stream flow and on the provision of a suitable environment for valuable fish and wild life. The protective influences of the forest are many and far-reaching and under certain conditions may be more important than the supplying of wood.

This is particularly true in some parts of the tropics and subtropics, in crucial drainage areas and in regions of light, sandy soil unsuitable for cultivation, where the removal of the forest cover often results in near-desert conditions. Forestry may also aim at the full development of the recreational value of an area; but even where recreation or protection are the main objectives, controlled cutting of wood can usually be carried on advantageously, if the cutting operations do not interfere with the primary purpose of management.

The term "multiple use" is widely used in referring to the management of a forest area for two or more purposes, one of which may be dominant, but not exclusive.

### Bases of Forestry

Forestry may be defined as the scientific management of forests for the continuous production of goods and services. While its practice involves many problems of an economic and administrative nature, the solution of which is essential for success, its proper development must in the first place be based on a sound knowledge of the forests and of the laws that govern their development and on an understanding of the characteristics of the individual species that make up the forests and of the interrelationship of the forests with their environment.

The forests of the world include a great many species of trees, not necessarily having any close botanical relationship, with varying adaptations to climate, soil and topography and more or less distinct reproductive characteristics, forms and rates of growth, sizes and ages at maturity, light requirements and abilities to resist damage by insects and other destructive agencies. It is with an intimate knowledge of these features of the trees and tree communities that sound forestry practice must begin.

## BOTANICAL RELATIONSHIPS

Trees represent many different families within the two major divisions of the spermatophytes or seed-bearing plants — the gymnosperms and the angiosperms. Trees of the gymnosperm group are more commonly known as conifers or softwoods, while those of the angiosperm group are often called broadleaf trees or hardwoods. These names do not apply universally — the wood of some gymnosperms, for example, is considerably harder than that of certain angiosperms. For most practical purposes, however, the forests of the world are classified as softwoods, temperate hardwoods and tropical hardwoods. The dominance of the softwoods in the northern temperate forests is of special economic significance, because man has found that these woods are particularly useful. Thus the saw timber that is used throughout the world in general building and construction is 75 per cent softwood and an even larger proportion of wood pulp comes from this class of trees. bole or trunk. Throughout the life of the tree there are alternate periods of growth and dormancy and in general, owing to differences in the structure of the wood produced at the beginning and at the end of a growing season, the successive layers appear as a series of concentric circles when the tree is cut transversely. Where climatic conditions are such that a period of growth alternates with a period of dormancy within the year, these woody layers are known as annual rings. The yearly increase in wood volume depends primarily on diameter growth.

For the first few years after its formation the wood, then known as "sapwood," contains some living cells which store food substances and conduct water and solutions of mineral nutrients from the roots to the crown of the tree. Later the cells die, other physiological changes take place and the wood, at this stage known as "heartwood," ceases to function except in support of the crown. Quite commonly the heartwood is darker in colour, more durable and has better timber properties than the sapwood.

The physical structure and consequently the density and technical properties of the wood laid down by the cambium, vary considerably from one species to another. The rate of diameter growth may also have diverse influences on the strength and quality of the wood: while rapid growth increases the strength of certain hardwoods, it may produce soft, spongy, Jow-grade timber in some of the softwoods.

The form of the bole generally varies according to the strength required to support the crown and resist wind pressures: in a closed forest the bole approaches a cylindrical form, whereas in exposed sites it is more conical owing to the need for mechanical support. There are also considerable differences between species, especially between the two major groups of hardwoods and softwoods, in the form of the crown and bole and in the vigour of development and persistence of side branches.

Usually softwoods produce a well-defined bole with relatively small branches, while the branches of many of the hardwoods are larger and more widely spreading. Since branches produce knots, for the production of clear timber it is desirable to remove the lower dead and dying branches while they are fairly small. In general, trees growing in a closed forest have fewer and smaller branches than those growing in the open. However, the rate at which lower branches are killed by shade and subsequently drop to the ground varies with the species.

Important characteristics of the mature root system also vary from one species to another. Some species have deeply penetrating tap roots. Others have mainly shallow lateral roots, which are more likely to be sensitive to drought and ground fires and to render the trees more susceptible to wind throw. Root modifications or adaptations that are related to the nutritional efficiency of certain species include:

- The presence of nodules containing nitrogen-fixing bacteria in the roots, particularly of alders and members of the Leguminosae,
- A close association between the root tips of many tree species and the mycelia of certain fungi: this association is known as "mycorrhiza."

## NUTRITIONAL AND ECOLOGICAL REQUIREMENTS

For its growth and for other life processes, the tree requires food and like other green plants it builds up its food materials from simple substances. The chemical constituents of a tree are chiefly carbon, hydrogen and oxygen, together with some nitrogen and certain minerals such as potash, lime, iron and phosphorus. The carbon is obtained from the small amount of carbon dioxide gas that is always present in the atmosphere and enters the plant through small openings in the leaf surface known as "stomata." Oxygen and hydrogen are obtained from the soil through the root system in the form of water carrying nitrogen (nitrates) and mineral nutrients in solution. The synthesizing of these elements to form substances that the tree requires -in particular the combination of carbon, hydrogen and oxygen to form carbohydrates — takes place mainly in the leaves in the presence of the characteristically green matter (chlorophyll) and of sunlight, which supplies the necessary energy. Air, water, chlorophyll and light are therefore essential to the production of the carbohydrates-in particular cellulose and lignin -which form by far the greater portion of the tree. Soil moisture is especially important, as a deficiency of water inhibits photosynthesis. Tree species differ markedly in their optimum water requirements and also in their sensitivity to departures from the optimum.

Tree species also differ greatly in the amount of light they require, or, alternatively, in the degree of shade they will tolerate. It is customary to classify species as shade-enduring or tolerant (of shade) and light-demanding or intolerant (of shade), but for most species the younger the tree and the better the soil quality on which it is growing, the greater its tolerance of shade.

The environmental factors important in tree growth — solar radiation, air temperature, humidity, precipitation and the physical and chemical nature of the soil — are usually spoken of as the factors of site or locality and their total effect with reference to a species as the site quality. Site quality is evaluated either directly by examination and classification of soil and other factors, or indirectly, as for example, by using ground vegetation as an indicator of soil conditions, or by relating the height of dominant trees to the age of the stand.

Since the bulk of the tree is derived from carbon dioxide and water, a relatively small demand is made on the soil in growing a crop of timber; thus land too poor for profitable cultivation may be capable of growing a crop of timber satisfactorily. Deeply penetrating roots bring up from the deeper soil strata mineral nutrients that would not otherwise be available; a relatively high proportion of these minerals may be deposited annually on the surface soil through the falling of leaves, twigs and other debris.

The leaves, which play such an important part in the nutrition of the tree, function for only a limited time before dying. Deciduous trees lose all their leaves at the end of the growing season, whereas evergreens lose only a portion of the older leaves each year.

The growth and general well-being of a tree depend primarily on soil and weather conditions. Also of vital importance to the future of each member of the forest community is the intense competition that develops for space and food requirements-a struggle that inevitably kills many of the competitors and brings about significant changes in the environment of the survivors.

## LONGEVITY

Under favorable conditions a tree usually grows vigorously until it reaches maturity, after which the growth rate gradually slows down. Overmature trees grow little and show signs of decadence. The normal life expectancy of trees varies widely with the species: some temperate and tropical hardwoods reach maturity in less than 100 years and some softwoods in 100 to 200 years, but many species of both softwoods and hardwoods require 200 to 500 years to mature. On the Pacific coast of North America, ages of 500 to 700 years are common; the giant sequoia in California has been known to live for over 3,000 years, an example of extreme longevity.

## RESISTANCE TO DAMAGE

Trees, like other organisms, are damaged and destroyed by various agencies, including fire, insects, fungi, grazing animals, industrial fumes and such climatic factors as wind, frost and snow. The capacity to resist such damage depends upon the vitality and the protective devices of the species. A knowledge of the relative resistance of different tree species is of importance in forest management.

## DISTRIBUTION

Apart from the chance results of geological history, the continental distribution of tree species and the limits of tree growth, both in altitude and latitude, depend primarily on heat; the east-west distribution, on the other hand, is more often determined by moisture conditions—the rainfall in the interior of a continent may be too scanty to support tree growth. Thus each species of tree has a range of natural distribution determined primarily by climate. Obviously there is a great deal of overlapping and many species grow under similar conditions — in fact it is common to have a number of species sharing the same area in a mixed stand. Pure stands of a single species do not commonly occur in the natural forest, though stands approaching this condition may be found where a species is well adapted to extreme conditions of climate or soil moisture, or where fire has brought about a situation particularly favorable to one species. The most heterogeneous and complex forest communities are found in tropical regions of high rainfall.

The distribution of a species within its natural range is determined mainly by soil, topography and competing vegetation, as well as by local variations in climate, which also affect the long-term development of many species: climatic races or strains develop, more or less adapted to local variations of climate within the range as a whole. This fact accounts, in part at least, for the wide variation in results obtained by using seed of other than local origin in establishing forests.

Scots pine, for example, has developed a number of climatic races and experience in both Europe and North America shows that efforts to introduce a "foreign" race of this species without a proper correlation of climatic conditions at seed source and planting site will almost certainly prove unsatisfactory, particularly in the susceptibility of the trees to damage by snow and sleet and the incidence of disease. This, of course, also holds true for attempts to establish any tree as an exotic outside the boundaries of its natural range. However, where the climatic conditions of two regions are similar, the introduction of tree species from the one to the other may be undertaken successfully, as, for example, in the introduction of conifers from the Pacific coast of North America into Great Britain.

An understanding of the various biological factors discussed above is of fundamental importance in the development of forestry practice, particularly in the fields of silviculture and protection. It must be emphasized, however, that practical forest management must depend also on the knowledge and application of sound business principles and on the development of suitable methods and techniques to be used in collecting, analyzing and organizing the basic data to be obtained from the forest.

## Branches of Forestry

As is implied by what has been said above, forestry has a number of subdivisions, each of which is concerned with some particular aspect or phase of the subject. Hence, an understanding of forestry itself is dependent on knowledge of its parts, the extent to which they are interdependent and how best they may be co-ordinated.

## FOREST PROTECTION

Agencies that seriously damage and may destroy forests include fire, insects, diseases, grazing animals and adverse conditions of exposure and climate. The relative importance of these varies with local conditions. Fire is perhaps the most spectacular. It frequently destroys not only the forest growth, but also the surface organic layer beneath, thus reducing soil fertility. Fire is most serious in the coniferous forests of temperate regions and in the dry forests of the tropics and subtropics and is of course most difficult to control in heavily forested regions with sparse populations and poor accessibility.

Protection against fire is usually one of the first steps taken in the development of a forestry programme. Appropriate legislation is passed and organized staff and equipment and transportation facilities are provided to detect and suppress fires as they occur. Where the country is heavily and continuously wooded, strips may be cut through the forest to break the continuity of tree growth and delay or prevent the spread of fires. Clear-cut strips of this nature are known as "fireguards." Fire itself is sometimes used as a preventive measure where the annual fall of litter and the dying back of ground vegetation produce large accumulations of inflammable debris. The "early burning" of this material when moisture conditions are such as to prevent too rapid spread of fire will reduce the danger during a dry period later in the season.

Recent important advances in the field of fire protection include the use of aircraft, mainly for detection and of meteorological data to appraise more precisely the existing degree of fire hazard, so that dangerous conditions can be quickly recognized and adequate preventive measures taken.

In some European countries forest fires are rare and such as do occur cause negligible damage. In Norway, for example, forest properties may be insured against fire at a low premium rate. There are a number of reasons for this: *(a)* in general the forests are easily accessible and under intensive management, with annual fellings well distributed in a large number of small scattered holdings; *(b)* the forest consciousness of the people is highly developed; *(c)* the climatic conditions tend to be favorable, with well-distributed summer rains and an absence of prolonged hot, dry periods.

Despite the damage it causes, fire cannot be looked upon solely as an enemy of the forest. Under certain conditions controlled burning may prove a useful tool in securing the regeneration of valuable forest trees; it may prepare a suitable seedbed, eliminate competing vegetation, or assist in the distribution of seed. Certain conifers, for example, have serotinous cones and considerable heat is required to open them and release the seed: these species may reproduce abundantly after fire and the resulting stand is often referred to as a "fire type." Extensive even-aged coniferous forests of the northern temperate region owe their origin to fire, as do many other commercially important forests in various parts of the world.

Insects that damage and destroy trees and the fungi and other organisms that cause tree diseases, are present in all forests. In endemic proportions and in balance with their environment, they are a necessary and sometimes innocuous feature of the forest community. Even under these normal conditions, however, some insects and diseases may seriously damage commercially valuable trees. While the damage caused by fungi is usually slow in developing and of minor account in young trees, it is cumulative and may be so extensive in overmature trees as to render them valueless.

Destructive organisms may also reach *epidemic* proportions and destroy one or more tree species over wide areas. This is particularly likely to happen in overmature forests or when an organism is moved from its natural environment to new surroundings. A classical example is the chestnut blight *(Endothia parasitica),* which was brought to North America from Asia about 1900 and killed practically all of the sweet chestnut *(Castanea dentata)* in eastern America within a few years. Instances may also be cited of widespread destruction by forest insects, such as the spruce budworm *(Choristoneura fumiferana)* in the spruce and balsam-fir forests of northern and eastern Canada.

Methods used to control insects and disease include the removal of trees most susceptible to attack, the establishment of new stands containing a mixture of species rather than pure stands and discrimination against species susceptible to an insect or disease known to be present in the locality. For example, white pine *(Pinus strobus),* which is often badly attacked by the pine weevil (Pissodes strobi) and the blister rust *(Cronartium ribicola),* should not be encouraged where these pests are commonly found, unless some effective means of control is available.

Other methods of controlling forest pests include the introduction of parasites and predators, the use of chemical sprays and the development through selection and crossbreeding of strains or hybrids that are relatively immune to attack by specific pests.

Domestic animals grazing in forest areas, particularly cattle and goats, often cause much damage by killing seedlings and compacting the soil. In the eastern Mediterranean region, where goats have been allowed to roam over the countryside, much of the forest vegetation has been completely destroyed. In many cases the only control possible is to exclude the animals; elsewhere forest grazing may be permitted under carefully controlled conditions.

Forests are normally protected against the adverse effects of weather and exposure by some adjustment in the methods of cutting and by encouraging the development of species best suited to local conditions.

# 7

# Effects of Forest Certification on Biodiversity

## POPULATION GROWTH, VARIATION AMONG NATIONS

Our global human population, 6 billion at present, will cross the 7 billion mark by 2015. The needs of this huge number of human beings cannot be supported by the Earth's natural resources, without degrading the quality of human life. In the near future, fossil fuel from oil fields will run dry. It will be impossible to meet the demands for food from existing agro systems. Pastures will be overgrazed by domestic animals and industrial growth will create ever-greater problems due to pollution of soil, water and air.

Seas will not have enough fish. Larger ozone holes will develop due to the discharge of industrial chemicals into the atmosphere, which will affect human health. Global warming due to industrial gases will lead to a rise in sea levels and flood all low-lying areas, submerging coastal agriculture as well as towns and cities. Water 'famines' due to the depletion of fresh water, will create unrest and eventually make countries go to war. The control over regional biological diversity, which is vital for producing new medicinal and industrial products, will lead to grave economic conflicts between biotechnologically advanced nations and the biorich countries.

Degradation of ecosystems will lead to extinction of thousands of species, destabilizing natural ecosystems of great value. These are only some of the environmental problems related to an increasing human population and more intensive use of resources that we are likely to face in future. These effects can be averted by creating a mass environmental awareness movement that will bring about a change in people's way of life.

Increase in production per capita of agricultural produce at a global level ceased during the 1980's. In some countries, food shortage has become a permanent feature. Two of every three children in South Africa are underweight. In other regions famines due to drought have become more frequent. Present development strategies have not been able to successfully address these problems related to hunger and malnutrition. On the other hand, only 15 per cent of the world's population in the developed world is earning 79 per cent of income! Thus the disparity in the extent of per capita resources that are used by people who live in a 'developed' country as against those who live in a 'developing' country is extremely large. Similarly, the disparity between the rich and the poor in India is also growing.

The increasing pressures on resources place great demands on the in-built buffering action of nature that has a certain ability to maintain a balance in our environment. However, current development strategies that essentially lead to short-term gains have led to a breakdown of our Earth's ability to replenish the resources on which we depend.

## GLOBAL POPULATION GROWTH

The world population is growing by more than 90 million per year, of which 93 per cent is in developing countries. This will essentially prevent their further economic 'development'. In the past, population growth was a gradual phenomenon and the Earth's ability to replenish resources was capable of adjusting to this increase. In the recent past, the escalation in growth of human numbers has become a major cause of our environmental problems. Present projections show that if our population growth is controlled, it will still grow to 7.27 billion by 2015.

However, if no action is taken it will become a staggering 7.92 billion. Human population growth increased from: 1 to 2 billion, in 123 years. 2 to 3 billion, in 33 years. 3 to 4 billion, in 14 years. 4 to 5 billion, in 13 years. 5 to 6 billion, in 11 years. It is not the census figures alone that need to be stressed, but an appreciation of the impact on natural resources of the rapid escalation in the rate of increase of human population in the recent past. The extent of this depletion is further increased by affluent societies that consume per capita more energy and resources, that less fortunate people. This is of great relevance for developing a new ethic for a more equitable distribution of resources. In the first half of the 1900s human numbers were growing rapidly in most developing countries such as India and China. In some African countries the growth was also significant. In contrast, in the developed world population growth had slowed down. It was appreciated that the global growth rate was depleting the Earth's resources and was a direct impediment to human development.

Several environmental ill-effects were linked with the increasing population of the developing world. Poverty alleviation programmes failed, as whatever was done was never enough as more and more people had to be supported on Earth's limited resources. In rural areas population growth led to increased fragmentation of farm land and unemployment. In the urban sector it led to inadequate housing and an increasing level of air pollution from traffic, water pollution from sewage, and an inability to handle solid waste. By the 1970s most countries in the developing world had realised that if they had to develop their economics and improve the lives of their citizens they would have to curtail population growth.

Though population growth shows a general global decline, there are variations in the rate of decline in different countries. By the 1990s the growth rate was decreasing in most countries such as China and India. The decline in the 90s was greatest in India. However, fertility continues to remain high in sub Saharan African countries. There are cultural, economic, political and demographic reasons that explain the differences in the rate of population control in different countries. It also varies in different parts of certain countries and is linked with community and/or religious thinking. Lack of Government initiatives for Family Welfare Programme and a limited access to a full range of contraceptive measures are serious impediments to limiting population growth in several countries.

## POPULATION EXPLOSION-FAMILY WELFARE PROGRAMME

In response to our phenomenal population growth, India seriously took up an effective Family Planning Programme which was renamed the Family Welfare Programme. Slogans such as 'Hum do hamare do' indicated that each family should not have more than two children. It however has taken several decades to become effective. At the global level by the year 2000,600 million, or 57 per cent of women in the reproductive age group, were using some method of contraception.

However the use of contraceptive measures is higher in developed countries–68 per cent, and lower in developing countries-55 per cent. Female sterilization is the most popular method of contraception used in developing countries at present. This is followed by the use of oral contraceptive pills and, intrauterine devices for women, and the use of condoms for men. India and China have been using permanent sterilization more effectively than many other countries in the developing world. The best decision for the method used by a couple depends on a choice that they make for themselves.

This must be based on good advice from doctors or trained social workers who can suggest the full range of methods available for them to choose from. Informing the public about the various contraceptive measures that are available is of primary importance. This must be done actively by Government Agencies such as Health and Family Welfare, as well as Education and Extension workers. It is of great importance for policy makers and elected representatives of the people–Ministers, MPs, MLAs at Central and State levels–to understand the great and urgent need to support Family Welfare.

The media must keep people informed about the need to limit family size and the ill effects of a growing population on the worlds resources. The decision to limit family size depends on a couple's background and education. This is related to Government Policy, the effectiveness of Family Welfare Programmes, the educational level, and information levels in mass communication. Free access to Family Welfare information provided through the Health Care System, is in some cases unfortunately counteracted by cultural attitudes.

Frequently misinformation and inadequate information are reasons why a family does not go in for limiting its size. The greatest challenge the world now faces is how to supply its exploding human population with the resources it needs. It is evident that without controlling human numbers, the Earth's resources will be rapidly exhausted. In addition economically advanced countries and rich people in poorer countries use up more resources than they need.

As population expands further, water shortages will become acute. Soil will become unproductive. Rivers, lakes and coastal waters will be increasingly polluted. Water related diseases already kill 12 million people every year in the developing world. By 2025, there will be 48 countries that are starved for water. Air will become increasingly polluted. Air pollution already kills 3 million people every year. The first 'green revolution' in the '60s produced a large amount of food but has led to several environmental problems. Now, a new green revolution is needed, to provide enough food for our growing population, that will not damage land, kill rivers by building large dams, or spread at the cost of critically important forests, grasslands and wetlands. The world's most populous regions are in coastal areas.

These are critical ecosystems and are being rapidly destroyed. Global climate change is now a threat that can affect the very survival of high population density coastal communities. In the sea, fish populations are suffering from excessive fishing. Once considered an inexhaustible resource, over fishing has depleted stocks extremely rapidly. It will be impossible to support further growth in coastal populations on existing fish reserves. Human populations will inevitably expand from farm lands into the remaining adjacent forests. Many such encroachments in India have been regularised over the last few decades.

But forest loss has long-term negative effects on water and air quality and the loss of biodiversity is still not generally seen as a major deterrent to human well-being. The extinction of plant and animal species resulting from shrinking habitats threatens to destroy the Earth's living web of life. Energy use is growing, both due to an increasing population, and a more energy hungry lifestyle that increasingly uses consumer goods that require large amounts of energy for their production, packaging, and transport. Our growing population also adds to the enormous amount of waste. With all these linkages between population growth and the environment, Family Welfare Programmes have become critical to human existence.

## PLANNING FOR THE FUTURE

How Governments and people from every community meet challenges such as limiting population size, protecting the natural environment, change their consumer oriented attitudes, reduce habits that create excessive waste, elevates poverty and creates an effective balance between conservation and development will determine the worlds future.

## THE URBAN CHALLENGE

Population increases will continue in urban centers in the near future. The UN has shown that by 2025 there will be 21 "megacities" most of which will be situated in developing countries. Urban centers are already unable to provide adequate housing, services such as water and drainage systems, growing energy needs, or better opportunities for income generation.

## METHODS OF STERILIZATION

India's Family Welfare Programme has been fairly successful but much still needs to be achieved to stabilize our population. The most effective measure is the one most suited to the couple once they have been offered all the various options that are available. The Family Welfare Programme advocates a variety of measures to control population.

Permanent methods or sterilisation are done by a minor surgery. Tubectomy in females is done by tying the tubes that carry the ovum to the uterus. Male sterilization or vasectomy, is done by tying the tubes that carry the sperm. Both are very simple procedures, done under local anesthesia, are painless and patients have no post operative problems.

Vasectomy does not cause any loss in the male's sexual ability but only arrests the discharge of sperm. There are several methods of temporary birth control. Condoms are used by males to prevent sperms from fertilizing the ovum during intercourse. Intrauterine devices (Copper Ts) are small objects which can be placed by a doctor in the uterus so that the ovum cannot be implanted, even if fertilized. They do not disturb any functions in the woman's life or work. Oral contraceptive tablets (pills) and injectable drugs are available that prevent sperms from fertilizing the ovum. There are also traditional but less reliable methods of contraception such as abstinence of the sexual act during the fertile period of the women's cycle and withdrawal during the sexual act.

## URBANIZATION

In 1975 only 27 per cent of the people in the developing world lived in urban areas. By 2000 this had grown to 40 per cent and by 2030 well informed estimates state that this will grow to 56 per cent. The developed world is already highly urbanized with 75 per cent of its population living in the urban sector.

## URBAN ENVIRONMENTS

Nearly half the world's population now lives in urban areas. The high population density in these areas leads to serious environmental issues. Today, more than 290 million people live in towns and cities in India. There were 23 metros in India in 1991, which grew to 40 by 2001. Urban population growth is both due to migration of people to towns and cities from the rural sector in search of better job options as well as population growth within the city.

As a town grows into a city it not only spreads outwards into the surrounding agricultural land or natural areas such as forests, grasslands and wetlands but also grows skywards with high rise buildings. The town also loses its open spaces and green cover unless these are consciously preserved. This destroys the quality of life in the urban area. Good urban planning is essential for rational landuse planning, for upgrading slum areas, improving water supply and drainage systems, providing adequate sanitation, developing effective waste water treatment plants and an efficient public transport system.

Unplanned and haphazard growth of urban complexes has serious environmental impacts. Increasing solid waste, improper garbage disposal and air and water pollution are frequent side effects of urban expansions.

While all these issues appear to be under the preview of local Municipal Corporations, better living conditions can only become a reality if every citizen plays an active role in managing the environment. This includes a variety of "Dos and Don'ts" that should become an integral part of our personal lives.

Apart from undertaking actions that support the environment every urban individual has the ability to influence a city's management. He or she must see that the city's natural green spaces, parks and gardens are maintained, river and water fronts are managed appropriately, roadside tree cover is maintained, hill slopes are afforested and used as open spaces and architectural and heritage sites are protected. Failure to do this leads to increasing urban problems which eventually destroys a city's ability to maintain a healthy and happy lifestyle for its dwellers.

All these aspects are closely linked to the population growth in the urban sector. In many cities growth outstrips the planner's ability to respond to this in time for a variety of reasons.

| Mega cities in India | Population (in millions) in 2001 | Projection (in millions) for 2015 |
|---|---|---|
| Mumbai | 16.5 | 22.6 |
| Kolkata | 13.3 | 16.7 |
| Delhi | 13.0 | 20.9 |

Small urban centers too will grow rapidly during the next decades and several rural areas will require reclassification as urban centers. India's urban areas will grow by a projected 297 million residents. In India people move to cities from rural areas in the hope of getting a better income. This is the 'Pull' factor. Poor opportunities in the rural sector thus stimulates migration to cities. Loss of agricultural land to urbanisation and industry, the inability of governments to sustainably develop the rural sector, and a lack of supporting infrastructure in rural areas, all push people from the agricultural and natural wilderness ecosystems into the urban sector. As our development strategies have focused attention mostly on rapid industrial development and relatively few development options are offered for the agricultural rural sector, a shift of population is inevitable.

As population in urban centers grows, they draw on resources from more and more distant areas. The "Ecological footprint" corresponds to the land area necessary to supply natural resources and disposal of waste of a community. At present the average ecological footprint of an individual at the global level is said to be 2.3 hectares of land per capita. But it is estimated that the world has only 1.7 hectares of land per individual to manage these needs sustainably.

This is thus an unsustainable use of land. The pull factor of the urban centers is not only due to better job opportunities, but also better education, health care and relatively higher living standards. During the last few decades in India, improvements in the supply of clean water, sanitation, waste management, education and health care has all been urban centric, even though the stated policy has been to support rural development.

Thus in reality, development has lagged behind in the rural sector that is rapidly expanding in numbers. For people living in wilderness areas in our forests and mountain regions, development has been most neglected. It is not appropriate to use the development methods used for other rural communities for tribal people who are dependent on collecting natural resources from the forests.

A different pattern of development that is based on the sustainable extraction of resources from their own surroundings would satisfy their development aspirations. In general the growing human population in the rural sector will only opt to live where they are if they are given an equally satisfying lifestyle.

## THE WILDERNESS–RURAL-URBAN LINKAGE

The environmental stresses caused by urban individuals covers an 'ecological footprint' that goes far beyond what one expects. The urban sector affects the land at the fringes of the urban area and the areas from which the urban center pulls in agricultural and natural resources.

Urban centers occupy 2 per cent of the worlds' land but use 75 per cent of the industrial wood. About 60 per cent of the world's water is used by urban areas of which half irrigates food crops for urban dwellers, and one third goes to industry and the rest is used for household use and drinking water.

The impact that urban dwellers have on the environment is not obvious to them as it happens at distant places which supports the urban ecosystem with resources from agricultural and even more remote wilderness ecosystems.

## URBAN POVERTY AND THE ENVIRONMENT

The number of poor people living in urban areas is rapidly increasing. A third of the poor people in the world live in urban centers. These people live in hutments in urban slums and suffer from water shortages and unsanitary conditions. In most cases while a slum invariably has unhygenic surroundings, the dwellings themselves are kept relatively clean.

It is the 'common' areas used by the community that lacks the infrastructure to maintain a hygienic environment. During the 1990s countries that have experienced an economic crisis have found that poor urban dwellers have lost their jobs due to decreasing demands for goods, while food prices have risen. Well paid and consistent jobs are not as easily available in the urban centers at present as in the past few decades. One billion urban people in the world live in inadequate housing, mostly in slum areas, the majority of which are temporary structures.

However, low income groups that live in high rise buildings can also have high densities and live in poor unhygienic conditions in certain areas of cities. Illegal slums often develop on Government land, along railway tracks, on hill slopes, riverbanks, marshes, etc. that are unsuitable for formal urban development. On the riverbanks floods can render these poor people homeless. Adequate legal housing for the urban poor remains a serious environmental concern. Urban poverty is even more serious than rural poverty, as unlike the rural sector, the urban poor have no direct access to natural resources such as relatively clean river water, fuelwood and non wood forest products. The urban poor can only depend on cash to buy the goods they need, while in the rural sector they can grow a substantial part of their own food. Living conditions for the urban poor are frequently worse than for rural poor.

Both outdoor and indoor air pollution due to high levels of particulate matter and sulphur dioxide from industrial and vehicle emissions lead to high death rates from respiratory diseases. Most efforts are targeted at outdoor air pollution. Indoor air pollution due to the use of fuel wood, waste material, coal, etc. in 'chulas' is a major health issue.

This can be reduced by using better designed 'smokeless' chulas, hoods and chimneys to remove indoor smoke. With the growing urban population, a new crisis of unimaginable proportions will develop in the next few years. Crime rates, terrorism, unemployment, and serious environmental health related issues can be expected to escalate. This can only be altered by stabilizing population growth on a war footing.

## ENVIRONMENT AND HUMAN HEALTH

Environment related issues that affect our health have been one of the most important triggers that have led to creating an increasing awareness of the need for better environmental management. Changes in our environment induced by human activities in nearly every sphere of life have had an influence on the pattern of our health. The assumption that human progress is through economic growth is not necessarily true.

We expect urbanization and industrialization to bring in prosperity, but on the down side, it leads to diseases related to overcrowding and an inadequate quality of drinking water, resulting in an increase in waterborne diseases such as infective diarrhoea and air borne bacterial diseases such as tuberculosis. High-density city traffic leads to an increase in respiratory diseases like asthma. Agricultural pesticides that enhanced food supplies during the green revolution have affected both the farm worker and all of us who consume the produce.

Modern medicine promised to solve many health problems, especially associated with infectious diseases through antibiotics, but bacteria found ways to develop resistant strains, frequently even changing their behaviour in the process, making it necessary to keep on creating newer antibiotics. Many drugs have been found to have serious side effects. At times the cure is as damaging as the disease process itself.

Thus development has created several long-term health problems. While better health care has led to longer life spans, coupled with a lowered infant mortality, it has also led to an unprecedented growth in our population which has negative implications on environmental quality. A better health status of society will bring about a better way of life only if it is coupled with stabilising population.

## ENVIRONMENTAL HEALTH

Environmental health, as defined by WHO, comprises those aspects of human health, including quality of life, that are determined by physical, chemical, biological, social, and psychosocial factors in the environment. It also refers to the theory and practice of assessing, correcting, controlling, and preventing those factors in the environment that adversely affect the health of present and future generations. Our environment affects health in a variety of ways.

Climate and weather affect human health. Public health depends on sufficient amounts of good quality food, safe drinking water, and adequate shelter. Natural disasters such as storms, hurricanes, and floods still kill many people every year. Unprecedented rainfall trigger epidemics of malaria and water borne diseases. Global climate change has serious health implications. Many countries will have to adapt to uncertain climatic conditions due to global warming. As our climate is changing, we may no longer know what to expect.

There are increasing storms in some countries, drought in others, and a temperature rise throughout the world. The El Niño winds affect weather worldwide. The El Niño event of 1997/98 had serious impacts on health and well-being of millions of people in many countries. It created serious drought, floods, and triggered epidemics. New strategies must be evolved to reduce vulnerability to climate variability and changes.

Economic inequality and environmental changes are closely connected to each other. Poor countries are unable to meet required emission standards to slow down climate change. The depletion of ozone in the stratosphere (middle atmosphere) also has an important impact on global climate and in turn human health, increasing the amount of harmful ultraviolet radiation that reaches the Earth's surface. This results in diseases such as skin cancer.

## BHOPAL GAS TRAGEDY

The siting of industry and relatively poor regulatory controls leads to ill health in the urban centers. Accidents such as the Bhopal gas tragedy in 1984 where Union Carbide's plant accidentally released 30 tones of methyl isocyanate, used in the manufacture of pesticides, led to 3,330 deaths and 1.5 lakh injuries to people living in the area. Development strategies that do not incorporate ecological safeguards often lead to ill health. Industrial development without pollution control and traffic congestion affect the level of air pollution in many cities.

On the other hand, development strategies that can promote health invariably also protect the environment. Thus environmental health and human health are closely interlinked. An improvement in health is central to sound environmental management. However this is rarely given sufficient importance in planning development strategies.

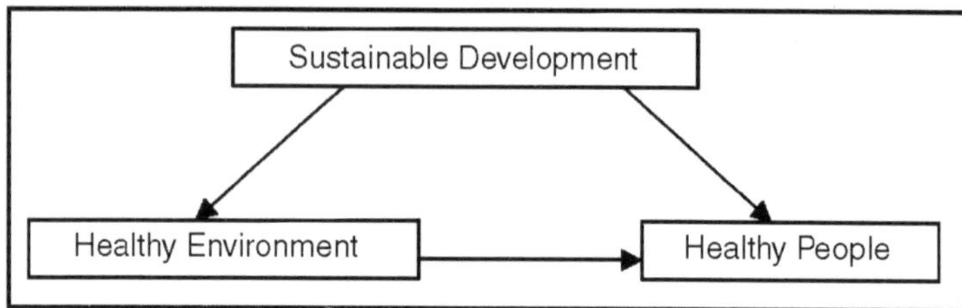


## EXAMPLES OF THE LINKAGES

- Millions of children die every year due to diarrhoea from contaminated water or food. An estimated 2000 million people are affected by these diseases and more than 3 million children die each year from waterborne diseases across the world. In India, it is estimated that every fifth child under the age of 5 dies due to diarrhoea. This is a result of inadequate environmental management and is mainly due to inadequate purification of drinking water. Wastewater and/or sewage entering water sources without being treated leads to continuous gastrointestinal diseases in the community and even sporadic large epidemics. Large numbers of people in tropical countries die of malaria every year and millions are infected. An inadequate environmental management of stagnant water, which forms breeding sites of Anopheles mosquitoes is the most important factor in the spread of malaria. The resurgence of malaria in India is leading to cerebral malaria that affects the brain and has a high mortality.

- Millions of people, mainly children, have poor health due to parasitic infections, such as amoebiasis and worms. This occurs from eating infected food, or using poor quality water for cooking food. It is estimated that 36 per cent of children in low-income countries and 12 per cent in middle income countries are malnourished. In India, about half the children under the age of four are malnourished and 30 per cent of newborns are significantly underweight.
- Hundreds of millions of people suffer serious respiratory diseases, including lung cancer and tuberculosis, from crowded homes and public places. Motor vehicle exhaust fumes, industrial fumes, tobacco smoke and cooking food on improper ‘chulas’, contribute to respiratory diseases.
- Millions of people are exposed to hazardous chemicals in their workplace or homes that lead to ill health due to industrial products where controls are not adhered to.
- Tens of thousands of people in the world die due to traffic accidents due to inadequate management of traffic conditions. Poor management at the accident site, and inability to reach a hospital within an hour causes a large number of deaths, especially from head injuries.
- Basic environmental needs such as clean water, clean air and adequate nutrition which are all related to environmental goods and services do not reach over 1000 million people living in abject poverty.
- Several million people live in inadequate shelters or have no roof over their heads especially in urban settings. This is related to high inequalities in the distribution of wealth and living space.
- Population growth and the way resources are being exploited and wasted, threatens environmental integrity and directly affects health of nearly every individual.
- Health is an outcome of the interactions between people and their environment. Better health can only come from a more sustainable management of the environment.

## IMPORTANT STRATEGIC CONCERNS

- The world must address people’s health care needs and sustainable use of natural resources, which are closely linked to each other.

- Strategies to provide clean pottable water and nutrition to all people is an important part of a healthy living environment.
- Providing clean energy sources that do not affect health is a key to reducing respiratory diseases.
- Reducing environmental consequences of industrial and other pollutants such as transport emissions can improve the status of health.
- Changing patterns of agriculture away from harmful pesticides, herbicides and insecticides which are injurious to the health of farmers and consumers by using alternatives such as Integrated Pest Management and non-toxic biopesticides can improve health of agricultural communities, as well as food consumers.
- Changing industrial systems into those that do not use or release toxic chemicals that affect the health of workers and people living in the vicinity of industries can improve health and environment.
- There is a need to change from using conventional energy from thermal power that pollutes air and nuclear power that can cause serious nuclear disasters to cleaner and safer sources such as solar, wind and ocean power that do not affect human health. Providing clean energy is an important factor that can lead to better health.
- The key factors are to control human population and consume less environmental goods and services which could lead to 'health for all'. Unsustainable use of resources by an ever growing population leads to unhealthy lives. Activities that go on wasting environmental goods and destroying its services by producing large quantities of non-degradable wastes, leads to health hazards.
- Poverty is closely related to health and is itself a consequence of improper environmental management. An inequitable sharing of natural resources and environmental goods and services, is linked to poor health. The world's consumption of non-renewable resources is concentrated in the developed countries. Rich countries consume 50 times more per capita than people in less developed countries. This means that developed countries also generate proportionately high quantities of waste material, which has serious health concerns.

## DEFINITION OF HEALTH IMPACT ASSESSMENT (HIA) BY WHO

Health impact assessment is a combination of procedures, methods and tools by which a policy, programme or project may be judged as to its potential effects on the health of a population, and the distribution of those effects within the population.

## CLIMATE AND HEALTH

Human civilizations have adapted mankind to live in a wide variety of climates. From the hot tropics to the cold arctic, in deserts, marshlands and in the high mountains. Both climate and weather have a powerful impact on human life and health issues. Natural disasters created by extremes of weather (heavy rains, floods, hurricanes) which occur over a short period of time, can severely affect health of a community. Poor people are more vulnerable to the health impacts of climate variability than the rich.

Of approximately 80,000 deaths which occur world-wide each year as a result of natural disasters about 95 per cent are in poor countries. In weather-triggered disasters hundreds of people and animals die, homes are destroyed, crops and other resources are lost. Public health infrastructure, such as sewage disposal systems, waste management, hospitals and roads are damaged. The cyclone in Orissa in 1999 caused 10,000 deaths. The total number of people affected was estimated at 10 to 15 million! Human physiology can adapt to changes in weather, within certain limits. However, marked short-term fluctuations in weather lead to serious health issues. Heat waves cause heat-related illness and death (*e.g.,* heat stroke). The elderly and persons with existing heart or respiratory diseases are more vulnerable. Heat waves in India in 1998 were associated with many deaths.

Climate plays an important role in vector-borne diseases transmitted by insects such as mosquitoes. These disease transmitters are sensitive to direct effects of climate such as temperature, rainfall patterns and wind. Climate affects their distribution and abundance through its effects on host plants and animals. Malaria transmission is particularly sensitive to weather and climate. Unusual weather conditions, for example a heavy downpour, can greatly increase the mosquito population and trigger an epidemic.

In the desert and at highland fringes of malarious areas, malaria transmission is unstable and the human population lacks inherent protective immunity. Thus, when weather conditions (rainfall and temperature) favour transmission, serious epidemics occur in such areas. Fluctuations in malaria over the years have been linked to changes in rainfall associated with the El Niño cycle.

## INFECTIOUS DISEASES

Many infectious diseases have re-emerged with a vengeance. Loss of effective control over diseases such as malaria and tuberculosis, have led to a return of these diseases decades after being kept under stringent control. Other diseases were not known to science earlier and seem to have suddenly hit our health and our lives during the last few decades. AIDS, due to the Human Immunodeficiency Virus (HIV) caused through sexual transmission and Severe Acute Respiratory Syndrome (SARS) are two such examples.

While these cannot be directly related to environmental change, they affect the environment in which we live by forcing a change in lifestyles and behaviour patterns. For example the SARS outbreak prevented people from several countries from traveling to other countries for months, severely affecting national economies, airline companies and the tourism industry. Why have infectious diseases that were related to our environment that were under control suddenly made a comeback? Diseases such as tuberculosis have been effectively treated with anti-tubercular drugs for decades.

These antibiotics are used to kill off the bacteria that causes the disease. However nature's evolutionary processes are capable of permitting bacteria to mutate by creating new genetically modified strains. Those that change in a way so that they are not affected by the routinely used antibiotics begin to spread rapidly. This leads to a reemergence of the disease. In the case of tuberculosis this has led to multi-drug resistant tuberculosis.

This is frequently related to HIV which reduces an individual's immunity to bacteria such as mycobacterium tuberculosis that causes tuberculosis. The newer broad-spectrum antibiotics, antiseptics, disinfectants, and vaccines once thought of as the complete answer to infectious diseases have thus failed to eradicate infectious diseases. Experts in fact now feel that these diseases will be the greatest killers in future and not diseases such as malignancy or heart disease.

While antibiotic resistance is a well-known phenomenon there are other reasons for the reemergence of diseases. Overcrowding due to the formation of slums in the urban setting leads to several health hazards, including easier spread of respiratory diseases. Inadequate drinking water quality and poor disposal of human waste due to absence of a closed sewage system and poor garbage management are all urban health issues.

This has led to a comeback of diseases such as cholera and an increased incidence of diarrhea and dysentery as well as infectious hepatitis (jaundice). With increasing global warming disease patterns will continue to change. Tropical diseases spread by vectors such as the mosquito will undoubtedly spread malaria further away from the equator.

Global warming will also change the distribution of dengue, yellow fever, encephalitis, etc. Warmer wetter climates could cause serious epidemics of diseases such as cholera. El Nino which causes periodic warming is likely to affect rodent populations. This could bring back diseases such as the plague. Globalisation and infectious disease Globalization is a world-wide process which includes the internationalization of communication, trade and economic organization.

It involves parallel changes such as rapid social, economic and political adjustments. Whilst globalization has the potential to enhance the lives and living standards of certain population groups, for poor and marginalized populations in both the non-formal as well as formal economic sectors of developing countries, globalization enhances economic inequalities.

## TUBERCULOSIS

Tuberculosis (TB) kills approximately 2 million people each year. In India the disease has reemerged and is now more difficult to treat. A global epidemic is spreading and becoming more lethal. The spread of HIV/AIDS and the emergence of multidrug-resistant tuberculosis is contributing to the increasing morbidity of this disease. In 1993, the World Health Organization (WHO) declared that tuberculosis had become a global emergency.

It is estimated that between 2002 and 2020, approximately 1000 million people will be newly infected, over 150 million people will get sick, and 36 million will die of TB–if its control is not rapidly strengthened. TB is a contagious disease that is spread through air. Only people who are sick with pulmonary TB are infectious. When infectious people cough, sneeze, talk or spit, they emit the tubercle bacilli into the air.

When a healthy person inhales these, he gets infected by the disease. Symptoms include prolonged fever, coughing spells and weight loss. It is estimated that, left untreated, each patient of active tuberculosis will infect on an average between 10 to 15 people every year. But people infected with TB will not necessarily get sick with the disease.

The immune system can cause the TB bacilli, which is protected by a thick waxy coat, to remain dormant for years. When an individual's immune system is weakened, the chances of getting active TB are greater.

- Nearly 1 per cent of the world's population is newly infected with TB each year.
- It is estimated that overall, one third of the world's population is likely to be infected with the tuberculosis bacillus at some point in time.
- Five to ten per cent of people who are infected with TB (but who are not infected with HIV) become sick or infectious at some time during their life. (WHO, 2002). Factors Contributing to the rise in tuberculosis
- TB kills about 2 million people each year (including persons infected with HIV).
- More than 8 million people become sick with TB each year, one person in the world every second!
- About 2 million TB cases per year occur in sub-Saharan Africa. This number is rising rapidly as a result of the HIV/AIDS epidemic.
- Around 3 million TB cases per year occur in South-east Asia.
- Over a quarter of a million TB cases per year occur in Eastern Europe.

## TUBERCULOSIS IN INDIA

There are 14 million TB patients in India, account for one third of the global cases of TB. Everyday 20,000 Indians contract TB and more than 1,000 die due to this chronic illness. TB attacks working adults in the age group of 15 to 50 years.

**HIV IS ACCELERATING THE SPREAD OF TB**

The link between HIV and TB affects a large number of people, each disease speeding the other's progress. HIV weakens the immune system. Someone who is HIV-positive and infected with TB is many times more likely to become seriously sick with TB rather than someone infected with TB who is HIV-negative. Tuberculosis is a leading cause of death among people who are HIV-positive, accounting for about 11 per cent of AIDS deaths worldwide.

**POORLY MANAGED TB PROGRAMMES ARE THREATENING TO MAKE TB INCURABLE**

Until 50 years ago, there were no drugs to cure tuberculosis. Now, strains that are resistant to one or more anti-TB drugs have emerged. Drugresistant tuberculosis is caused by inconsistent or partial treatment, when patients do not take all their drugs regularly for the required period, when doctors or health workers prescribe inadequate treatment regimens or where the drug supply is unreliable.

From a public health perspective, poorly supervised or incomplete treatment of TB is worse than no treatment at all. When people fail to complete standard treatment regimens, or are given the wrong treatment regimen, they may remain infectious. The bacilli in their lungs may develop resistance to anti-TB drugs. People they infect will have the same drug-resistant strain. While drug-resistant TB is treatable, it requires extensive chemotherapy that is often very expensive and is also more toxic to patients.

**MALARIA**

Malaria is a life-threatening parasitic disease transmitted by mosquitoes. The cause of malaria, a single celled parasite called plasmodium, was discovered in 1880.Later it was found that the parasite is transmitted from person to person through the bite of a female Anopheles mosquito, which requires blood for the growth of her eggs. Today approximately 40 per cent of the world's population, mostly those living in the world's poorest countries, risk getting malaria.

The disease was once more widespread but it was successfully eliminated from many countries with temperate climates during the mid 20th century. Today malaria has returned and is found throughout the tropical and sub-tropical regions of the world and causes more than 300 million acute illnesses and at least one million deaths annually (WHO). There are several types of human malaria.

Falciparum malaria is the most dangerous type of infection and is most common in Africa south of the Sahara, where it accounts for extremely high mortality rates. There are also indications of the spread of P. falciparum malaria in India and it has reappeared in areas where it had been eliminated. The malaria parasite enters the human host when an infected Anopheles mosquito bites an individual. Inside the human host, the parasite undergoes a series of changes as part of its complex life-cycle. Its various stages allow plasmodia to evade the immune system, infect the liver and red blood cells, and finally develop into a form that is able to infect a mosquito again when it bites an infected person.

Inside the mosquito, the parasite matures until it reaches the sexual stage where it can again infect a human host when the mosquito takes her next blood meal, 10 or more days later. Malaria symptoms appear about 9 to 14 days after the mosquito bite, although this varies with different plasmodium species.

Malaria produces high fever, headache, vomiting and body ache. If drugs are not available for treatment, or the parasites are resistant to them, the infection can progress rapidly to become life-threatening. Malaria can kill by infecting and destroying red blood cells (anaemia) and by clogging the capillaries that carry blood to the brain (cerebral malaria) or other vital organs.

Malaria parasites are developing unacceptable levels of resistance to drugs. Besides this, many insecticides are no longer useful against mosquitoes transmitting the disease. Good environmental management by clearing pools of stagnant water during the monsoons is effective in reducing the number of mosquitoes. Mosquito nets treated with insecticide reduce malaria transmission and child deaths.

Prevention of malaria in pregnant women, through measures such as Intermittent Preventive Treatment and the use of insecticide-treated nets (ITNs), results in improvement in maternal health, as well as infant health and survival. Prompt access to treatment with effective up-to-date medicines, such as artemisinin-based combination therapies (ACTs), saves lives. If countries can apply these and other measures on a wide scale and monitor them carefully, the burden of malaria on society will be significantly reduced.

## WATER-RELATED DISEASES

### WATER SUPPLY, SANITATION AND HYGIENE DEVELOPMENT

Among the main problems are a lack of priority given to this sector, lack of financial resources, erratic water supply and sanitation services, poor hygiene related behaviour patterns, and inadequate sanitation in public places such as schools, hotels, hospitals, health centers, etc. One of the most important aspects is a lack of environmental education and awareness that these disease processes are related to poor environment management in various sectors.

Providing access to sufficient quantities of safe water, the provision of facilities for a sanitary disposal of excreta, and introducing sound hygiene related behaviour can reduce the morbidity and mortality caused by these risk factors.

Environmental Sanitation and Hygiene Development About 2.4 billion people globally live under highly unsanitary conditions.

Poor hygiene and behaviour pattern increase the exposure to risk of incidence and spread of infectious diseases. Water improperly stored in homes is frequently contaminated by inadequate management at the household level. This can be easily reduced through education and awareness of how waterborne diseases are transmitted.

## HEALTH AND WATER RESOURCES DEVELOPMENT

An important aspect related to water-related diseases (in particular: water-related vectorborne diseases) is attributable to the way water resources are developed and managed. In many parts of the world the adverse health impacts of dam construction, irrigation development and flood control is related to increased incidence of malaria, Japanese encephalitis, schistosomiasis, lymphatic filariasis and other conditions. Other health issues indirectly associated with water resources development include nutritional status, exposure to agricultural pesticides and their residues.

## WATER BORNE DISEASES

Arid areas with rapidly expanding populations are already facing a crisis over water. Conservation of water and better management is an urgent need. The demand and supply balance is a vital part of developing sustainable use of water. This is being termed the 'Blue Revolution' and needs Governments, NGOs and people to work together towards a better water policy at International, National, State, regional and local levels. Locally good watershed management is a key to solving local rural problems. Present patterns of development are water hungry and water wasters. They do not address pollution and overuse. The linkages between managing water resources and health issues are have not been prioritised as a major source of environmental problems that require policy change, administrative capacity building and an increased financial support.

## THERE ARE 4 MAJOR TYPES OF WATER RELATED DISEASES

1. *Water borne diseases*: These are caused by dirty water contaminated by human and animal wastes, especially from urban sewage, or by chemical wastes from industry and agriculture. Some of these diseases, such as cholera and typhoid, cause serious epidemics. Diarrhoea, dysentery, polio, meningitis, and hepatitis A and E, are caused due to improper drinking water.

   Excessive levels of nitrates cause blood disorders when they pollute water sources. Pesticides entering drinking water in rural areas cause cancer, neurological diseases and infertility. Improving sanitation and providing treated drinking water reduces the incidence of these diseases.

2. *Water based diseases*: Aquatic organisms that live a part of their life cycle in water and another part as a parasite in man, lead to several diseases. In India, guinea worm affects the feet. Round worms live in the small intestine, especially of children.
3. *Water related vector diseases*: Insects such as mosquitoes that breed in stagnant water spread diseases such as malaria and filariasis. Malaria that was effectively controlled in India, has now come back as the mosquitoes have become resistant to insecticides. In addition, anti-malarial drugs are now unable to kill the parasites as they have become resistant to drugs.

   Change in climate is leading to the formation of new breeding sites. Other vector born diseases in India include dengue fever and filariasis. Dengue fever carries a high mortality. Filariasis leads to fever and chronic swelling over the legs.
4. *Water scarcity diseases:* In areas where water and sanitation is poor, there is a high incidence of diseases such as tuberculosis, leprosy, tetanus, etc. which occur when hands are not adequately washed.

## ARSENIC IN DRINKING WATER

Arsenic in drinking-water is a serious hazard to human health. It has attracted much attention since its recognition in the 1990s of its wide occurrence in wellwater in Bangladesh. It occurs less frequently in most other countries. The main source of arsenic in drinking water is arsenic-rich rocks through which the water has filtered. It may also occur because of mining or industrial activity in some areas. WHO has worked with other UN organizations to produce a state-of-the-art review on arsenic in drinking water.

Drinking water that is rich in arsenic leads to arsenic poisoning or arsenicosis. Excessive concentrations are known to occur in some areas. The health effects are generally delayed and the most effective preventive measure is supplying drinking water which is free of arsenic. Arsenic contamination of water is also due to industrial processes such as those involved in mining, metal refining, and timber treatment. Malnutrition may aggravate the effects of arsenic on blood vessels.

Water with high concentrations of arsenic if used over 5 to 20 years, results in problems such as colour changes on the skin, hard patches on the palms and soles, skin cancer, cancers of the bladder, kidney and lung, and diseases of the blood vessels of the legs and feet. It may also lead to diabetes, high blood pressure and reproductive disorders. Natural arsenic contamination occurs in Argentina, Bangladesh, Chile, China, India, Mexico, Thailand and the United

States. In China (in the Province of Taiwan) exposure to arsenic leads to gangrene, known as 'black foot disease'.

*Long term solutions for prevention of arsenicosis is based on providing safe drinking-water*:

- Deeper wells are often less likely to be contaminated.
- Testing of water for levels of arsenic and informing users.
- Monitoring by health workers-people need to be checked for early signs of arsenicosis-usually by observing skin problems in areas where arsenic in known to occur.
- Health education regarding harmful effects of arsenicosis and how to avoid them.

## ARSENIC POISONING–BANGLADESH

More than half the population of Bangladesh is threatened by high levels of arsenic found in drinking water. This could eventually lead to an epidemic of cancers and other fatal diseases. Rezaul Morol, a young Bangladeshi man, nearly died from arsenic poisoning caused by drinking arsenic-laden well-water for several years. The doctor advised Rezaul to stop drinking contaminated water and eat more protein-rich food such as fish. Since then Rezaul feels a lot better and is happy that his skin is healing.

## DIARRHOEA

Though several types of diarrhoea which give rise to loose motions and dehydration occur all over the world, this is especially frequently observed in developing countries. It causes 4 per cent of all deaths. In another 5 per cent it leads to loss of health. It is caused by gastrointestinal infections which kill around 2.2 million people globally each year. Most of these are children in developing countries. The use of contaminated water is an important cause of this group of conditions. Cholera and dysentery cause severe, sometimes life threatening and epidemic forms of these diseases.

## EFFECTS ON HEALTH

Diarrhoea is the frequent passage of loose or liquid stools. It is a symptom of various gastrointestinal infections. Depending on the type of infection, the diarrhoea may be watery (for example in cholera caused by vibrio cholera) or passed with blood and mucous (in dysentery caused by an amoeba, E Histolitica).

Depending on the type of infection, it may last a few days, or several weeks. Severe diarrhoea can become life threatening due to loss of excessive fluid and electrolytes such as Sodium and Potassium in watery diarrhoea.

This is particularly fatal in infants and young children. It is also dangerous in malnourished individuals and people with poor immunity. The impact of repeated diarrhoea on nutritional status is linked in a vicious cycle in children. Chemical or non-infectious intestinal conditions can also result in diarrhoea.

## CAUSES OF DIARRHOEA

Diarrhoea is caused by several bacterial, viral and parasitic organisms. They are mostly spread by contaminated water. It is more common when there is a shortage of clean water for drinking, cooking and cleaning. Basic hygiene is important in its prevention. Water contaminated with human feces surrounding a rural water source, or from municipal sewage, septic tanks and latrines in urban centers, are important factors in the spread of these diseases.

Feces of domestic animals also contain microorganisms that can cause diarrhoea through water. Diarrhoea is spread from one individual to another due to poor personal hygiene. Food is a major cause of diarrhoea when it is prepared or stored in unhygienic conditions. Water can contaminate food such as vegetables during irrigation. Fish and seafood from polluted water is a cause of severe diarrhoea. The infectious agents that cause diarrhoea are present in our environment.

In developed countries where good sanitation is available, most people get enough safe drinking water. Good personal and domestic hygiene prevents this disease which is predominantly seen in the developing world.

About 1 billion people do not have access to clean water sources and 2.4 billion have no basic sanitation. In Southeast Asia, diarrhoea is responsible for 8.5 per cent of all deaths. In 1998, diarrhoea was estimated to have killed 2.2 million people, most of whom were under 5 years of age.

*Interventions: Key measures to reduce the number of cases of diarrhoea include*:

- Access to safe drinking water.
- Improved sanitation.
- Good personal and food hygiene.
- Health education about how these infections spread.

*Key measures to treat diarrhoea include*:

- Giving more fluids than usual, (oral rehydration) with salt and sugar, to prevent dehydration.
- Continue feeding.
- Consulting a health worker if there are signs of dehydration or other problems.

In rural India, during the last decade public education through posters and other types of communication strategies has decreased infant mortality due to diarrhoea in several States. Posters depicting a child with diarrhoea being given water, salt and sugar solution to reduce death from dehydration has gone a long way in reducing both a serious condition requiring hospitalisation and intravenous fluids, as well as mortality.

## RISKS DUE TO CHEMICALS IN FOOD

Food contaminated by chemicals is a major worldwide public health concern. Contamination may occur through environmental pollution of the air, water and soil. Toxic metals, PCBs and dioxins, or the intentional use of various chemicals, such as pesticides, animal drugs and other agrochemicals have serious consequences on human health. Food additives and contaminants used during food manufacture and processing adversely affects health.

Diseases spread by food: Some foodborne diseases though well recognized, have recently become more common. For example, outbreaks of salmonellosis which have been reported for decades, has increased within the last 25 years. In the Western hemisphere and in Europe, Salmonella serotype Enteritidis (SE) has become a predominant strain. Investigations of SE outbreaks indicate that its emergence is largely related to consumption of poultry or eggs.

While cholera has devastated much of Asia and Africa for years, its reintroduction for the first time in almost a century on the South American continent in 1991 is an example of a well recognised infectious disease re-emerging in a region after decades. While cholera is often waterborne, many foods also transmit infection. In Latin America, ice and raw or underprocessed seafood are important causes for cholera transmission. Infection with a specific type of Escherichia coli (E. coli) was first described in 1982. Subsequently, it has emerged rapidly as a major cause of bloody diarrhoea and acute renal failure.

The infection is sometimes fatal, particularly in children. Outbreaks of infection, generally associated with beef, have been reported in Australia, Canada, Japan, United States, in various European countries, and in southern Africa.

Outbreaks have also implicated alfalfa sprouts, unpasteurized fruit juice, lettuce, game meat (meat of wild animals) and cheese curd. In 1996, an outbreak of Escherichia coli in Japan affected over 6,300 school children and resulted in 2 deaths. Listeria monocytogenes (Lm): The role of food in the transmission of this condition has been recognized recently.

In pregnant women, infections with Lm causes abortion and stillbirth. In infants and persons with a poor immune system it may lead to septicemia (blood poisoning) and meningitis. The disease is most often associated with consumption of foods such as soft cheese and processed meat products that are kept refrigerated for a long time, because Lm can grow at low temperatures.

Outbreaks of listeriosis have been reported from many countries, including Australia, Switzerland, France and the United States. Two recent outbreaks of Listeria monocytogenes in France in 2000 and in the USA in 1999 were caused by contaminated pork tongue and hot dogs respectively. Foodborne trematodes (worms) are increasing in South-east Asia and Latin America. This is related to a combination of intensive aquaculture production in unsanitary conditions, and consumption of raw or lightly processed fresh water fish and fishery products. Foodborne trematodes can cause acute liver disease, and may lead to liver cancer. It is estimated that 40 million people are affected worldwide.

Bovine Spongiform Encephalopathy (BSE), is a fatal, transmittable, neurodegenerative disease of cattle. It was first discovered in the United Kingdom in 1985. The cause of the disease was traced to an agent in sheep, which contaminated recycled bovine carcasses used to make meat and bone meal additives for cattle feed. Recycling of the BSE agent developed into a common source epidemic of more than 180,000 diseased animals in the UK alone.

The agent affects the brain and spinal cord of cattle which produces sponge-like changes visible under a microscope. About 19 countries have reported BSE cases and the disease is no longer confined to the European Community.

A case of BSE has been reported in a cattle herd in Japan. In human populations, exposure to the BSE agent (probably in contaminated bovine-based food products) has been strongly linked to the appearance in 1996 of a new transmissible spongiform encephalopathy of humans called variant Creutzfeldt-Jakob Disease (vCJD). By January 2002,119 people developed vCJD, most from the UK but five cases have been reported from France.

## CANCER AND ENVIRONMENT

Cancer is caused by the uncontrolled growth and spread of abnormal cells that may affect almost any tissue of the body. Lung, colon, rectal and stomach cancer are among the five most common cancers in the world for both men and women. Among men, lung and stomach cancer are the most common cancers worldwide. For women, the most common cancers are breast and cervical cancer.

In India, oral and pharangeal cancers form the most common type of cancer which are related to tobacco chewing. More than 10 million people are diagnosed with cancer in the world every year. It is estimated that there will be 15 million new cases every year by 2020.Cancer causes 6 million deaths every year–or 12 per cent of deaths worldwide. The causes of several cancers are known. Thus prevention of at least one-third of all cancers is possible.

Cancer is preventable by stopping smoking, providing healthy food and avoiding exposure to cancer-causing agents (carcinogens). Early detection and effective treatment is possible for a further one-third of cases. Most of the common cancers are curable by a combination of surgery, chemotherapy (drugs) or radiotherapy (X-rays). The chance of cure increases if cancer is detected early.

*Cancer control is based on the prevention and control of cancer by*:

- Promotion and strengthening of comprehensive national cancer control programmes.
- Building international networks and partnerships for cancer control.
- Promotion of organized, evidence based interventions for early detection of cervical and breast cancer.

- Development of guidelines on disease and programme management.
- Advocacy for a rational approach to effective treatments for potentially curable tumours.
- Support for low-cost approaches to respond to global needs for pain relief and palliative care.

## PREVENTION OF CANCER

Tobacco smoking is the single largest preventable cause of cancer in the world. It causes 80 to 90 per cent of all lung cancer deaths. Another 30 per cent of all cancer deaths, especially in developing countries include deaths from cancer of the oral cavity, larynx, oesophagus and stomach which are related to tobacco chewing. Preventive measures include bans on tobacco advertising and sponsorship, increased tax on tobacco products, and educational programmes which are undertaken to reduce tobacco consumption. Dietary modification is an important approach to cancer control.

Overweight individuals and obesity are known to be associated with cancer of the oesophagus, colon, rectum, breast, uterus and kidney. Fruit and vegetables may have a protective effect against many cancers. Excess consumption of red and preserved meat may be associated with an increased risk of colorectal cancer. Infectious agents are linked with 22 per cent of cancer deaths in developing countries and 6 per cent in industrialized countries.

Viral hepatitis B and C cause cancer of the liver. Human papilloma virus infection causes cancer of the cervix. The bacterium Helicobacter pylori increases the risk of stomach cancer. In some countries the parasitic infection schistosomiasis increases the risk of bladder cancer. Liver fluke increases the risk of cancer of the bile ducts. Preventive measures include vaccination and prevention of infection. Excessive solar ultraviolet radiation increases the risk of all types of cancer of the skin.

Avoiding excessive exposure to the sun, use of sunscreens and protective clothing are effective preventive measures. Asbestos is known to cause lung cancer. Aniline dyes have been linked to bladder cancer. Benzene can lead to leukaemia (blood cancer). The prevention of certain occupational and environmental exposure to several chemicals is an important element in preventing cancer.

## HUMAN RIGHTS

Several environmental issues are closely linked to human rights. These include the equitable distribution of environmental resources, the utilisation of resources and Intellectual Property Rights (IPRs), conflicts between people and wildlife especially around PAs, resettlement issues around development projects such as dams and mines, and access to health to prevent environment related diseases.

## EQUITY

One of the primary concerns in environmental issues is how wealth, resources and energy must be distributed in a community. We can think of the global community, regional community issues, national concerns and those related to a family or at the individual level. While economic disparities remain a fact of life, we as citizens of a community must appreciate that a widening gap between the rich and the poor, between men and women, or between the present and future generations must be minimised if social justice is to be achieved.

Today the difference between the economically developed world and the developing countries is unacceptably high. The access to a better lifestyle for men as against women is inherent in many cultures. Last but not the least, we in the present generation cannot greedily use up all our resources leaving future generations increasingly impoverished. Rights to land, water, food, housing are all a part of our environment that we all share. However, while some live unsustainable lifestyles with consumption patterns that the resource base cannot support, many others live well below the poverty line. Even in a developing country such as ours, there are enormous economic inequalities.

This requires an ethic in which an equitable distribution becomes a part of everyone's thinking. The people who live in the countries of the North and the rich from the countries in the South will have to take steps to reduce their resource use and the waste they generate.

Both the better off sectors of society and the less fortunate need to develop their own strategies of sustainable living and communities at each level must bring about more equitable patterns of wealth. The right to the use of natural resources that the environment holds is an essential component of human rights. It is related to disparities in the amount of resources available to different sectors of society.

People who live in wilderness communities are referred to as ecosystem people. They collect food, fuelwood, and non-wood products, fish in aquatic ecosystems, or hunt for food in forests and grasslands. When landuse patterns change from natural ecosystems to more intensively used farmland and pastureland the rights of these indigenous people are usually sacrificed. Take the case of subsidies given to the pulp and paper industry for bamboo which makes it several times cheaper for the industry than for a rural individual who uses it to build his home.

This infringes on the human right to collect resources they have traditionally used free of cost. Another issue is the rights of small traditional fishermen who have to contend against mechanised trawlers that impoverish their catch and overharvest fish in the marine environment. These people's right to a livelihood conflicts with the powerful economic interests of large-scale organised fisheries.

There are serious conflicts between the rights of rural communities for even basic resources such as water, and industrial development which requires large amounts of water for sustaining its productivity. The right to land or common property resources of tribal people is infringed upon by large development projects such as dams, mining and Protected Areas. Movements to protect the rights of indigenous peoples are growing worldwide. Reversing actions that have already been taken decades ago is a complex problem that has no simple solutions. In many cases a just tradeoff is at best achieved through careful and sensitively managed negotiations. This needs a deep appreciation of local environmental concerns as well as a sensitivity to the rights of local people.

## NUTRITION, HEALTH AND HUMAN RIGHTS

There are links between environment, nutrition and health which must be seen from a humanrights perspective. Proper nutrition and health are fundamental human rights. The right to life is a Fundamental Right in our constitution. As a deteriorating environment shortens life spans, this in effect has an impact on our fundamental constitutional right. Nutrition affects and defines the health status of all people, rich and poor.

It is linked to the way we grow, develop, work, play, resist infection and reach our aspirations as individuals, communities and societies. Malnutrition makes people more vulnerable to disease and premature death. Poverty is a major cause as well as a consequence of ill-health. Poverty, hunger, malnutrition and poorly managed environments together affect health and weaken the socioeconomic development of a country.

Nearly 30 per cent of humanity, especially those in developing countries– infants, children, adolescents, adults, and older persons are affected by this problem. A human rights approach is needed to appreciate and support millions of people left behind in the 20th century's health revolution. We must ensure that our environmental values and our vision are linked to human rights and create laws to support those that need a better environment, better health and a better lifestyle. Health and sustainable human development are equity issues. In our globalized 21st century, equity must begin at the bottom, hand in hand with a healthy environment, improved nutrition, and sustainable lifestyles. Putting first things first, we must also realise that resources allocated to preventing and eliminating disease will be effective only if the underlying causes such as malnutrition and environmental concerns, as well as their consequences, are successfully addressed.

## INTELLECTUAL PROPERTY RIGHTS AND COMMUNITY BIODIVERSITY REGISTERS

Traditional people, especially tribals living in forests, have used local plants and animals for generations. This storehouse of knowledge leads to many new 'discoveries' for modern pharmaceutical products. The revenue generated from such 'finds' goes to the pharmaceutical industry that has done the research and patented the product. This leaves the original tribal user with nothing while the industry could earn billions of rupees.

To protect the rights of indigenous people who have used these products, a possible tool is to create a Community Biodiversity Register of local products and their uses so that its exploitation by the pharmaceutical industry would have to pay a royalty to the local community. This however has still not been generally accepted. Mechanisms have to be worked out so that the local traditional users rights are protected.

Traditional Medicine: Traditional medicine refers to health practices, approaches, knowledge and beliefs that incorporate plant, animal and mineral based medicines, frequently of local or regional origin. It may be linked to spiritual therapies, manual techniques and exercises. These may be used singly or in combination to treat, diagnose and prevent illnesses or maintain well-being. Traditional medicine is often handed down through the generations or may be known to a special caste or tribal group.

Traditional medicine has maintained its popularity in all regions of the developing world and its use is rapidly spreading in industrialized countries. In India, some of our primary health care needs are taken care of entirely by traditional medicine, while in Africa, up to 80 per cent of the population uses it for primary health care. In industrialized countries, adaptations of traditional medicine are termed "Complementary" or "Alternative" Medicine (CAM). While there are advantages to traditional medicine as it is cheap and locally available, there are diseases which it cannot treat effectively.

This is a risk, as patients who use these alternative medicinal practices may rely on an ineffective measure. The consequences could be a serious delay in diagnosis and effective treatment of a treatable condition. There is a need to carefully research the claims of traditional practices to ensure that they are effective. In addition to patient safety issues, there is the risk that a growing herbal market and its great commercial benefit poses a threat to biodiversity through the over harvesting of the raw material for herbal medicines and other natural health care products.

This has been observed in the case of several Himalayan plants. If extraction from the wild is not controlled, this can lead to the extinction of endangered plant species and the destruction of natural habitats of several species. Another related issue is that at present, the requirements for protection provided under international standards for patent law and by most national conventional patent laws are inadequate to protect traditional knowledge and biodiversity. There are tried and tested scientific methods and products that have their origins in different traditional medicinal methods. Twenty-five per cent of modern medicines are made from plants first used traditionally. Yoga is known to reduce asthma attacks. Traditional Medicine has been found to be effective against several infectious diseases. Over one-third of the population in developing countries lack access to essential allopathic medicines. The provision of safe and effective TM/CAM therapies could become a tool to increase access to health care.

## VALUE EDUCATION

Value education in the context of our environment is expected to bring about a new sustainable way of life. Education both through formal and non-formal processes must thus address understanding environmental values, valuing nature and cultures, social justice, human heritage, equitable use of resources, managing common property resources and appreciating the cause of ecological degradation. Essentially, environmental values cannot be taught.

They are inculcated through a complex process of appreciating our environmental assets and experiencing the problems caused due to our destruction of our environment. The problems that are created by technology and economic growth are a result of our improper thinking on what 'development' means. Since we still put a high value only on economic growth, we have no concern for aspects such as sustainability or equitable use of resources. This mindset must change before concepts such as sustainable development can be acted upon.

Unsustainable development is a part of economic growth of the powerful while it makes the poor poorer. Consumerism is one aspect of this process favoured by the rich. As consumption of resources has till recently been an index of development, consumerism has thrived. It is only recently that the world has come to realise that there are other more important environmental values that are essential to bring about a better way of life.

Values in environment education must bring in several new concepts. Why and how can we use less resources and energy? Why do we need to keep our surroundings clean? Why should we use less fertilizers and pesticides in farms?

Why is it important for us to save water and keep our water sources clean? Or separate our garbage into degradable and non-degradable types before disposal? All these issues are linked to the quality of human life and go beyond simple economic growth. They deal with a love and respect for nature. These are the values that will bring about a better humanity, one in which we can live healthy, productive and happy lives in harmony with nature.

## ENVIRONMENTAL VALUES

Every human being has a great variety of feelings for different aspects of his or her surroundings. The Western, modern approach values the resources of Nature for their utilitarian importance alone. However true environmental values go beyond valuing a river for its water, a forest for its timber and non-wood forest products, or the sea for its fish.

Environmental values are inherent in feelings that bring about a sensitivity for preserving nature as a whole. This is a more spiritual, Eastern traditional value. There are several writings and sayings in Indian thought that support the concept of the oneness of all creation, of respecting and valuing all the different components of Nature. Our environmental values must translate to pro conservation actions in all our day to day activities. Most of our actions have adverse environmental impacts unless we consciously avoid them.

The sentiment that attempts to reverse these trends is enshrined in our environmental values. Values lead to a process of decision making which leads to action. For value education in relation to the environment, this process is learned through an understanding and appreciation of Nature's oneness and the importance of its conservation.

Humans have an inborn desire to explore Nature. Wanting to unravel its mysteries is a part of human nature. However, modern society and educational processes have invariably suppressed these innate sentiments. Once exposed to the wonders of the wilderness, people tend to bond closely to Nature.

They begin to appreciate its complexity and fragility and this awakens a new desire to want to protect our natural heritage. This feeling for Nature is a part of our Constitution, which strongly emphasises this value. Concepts of what constitutes right and wrong behaviour changes with time. Values are not constant.

It was once considered 'sport' to shoot animals. It was considered a royal, brave and much desirable activity to kill a tiger. In today's context, with wildlife reduced to a tiny fraction of what there was in the past, it is now looked down upon as a crime against biodiversty conservation. Thus the value system has been altered with time. Similarly with the large tracts of forest that existed in the past, cutting a few trees was not a significant criminal act.

Today this constitutes a major concern. We need a strong new environmental value system in which felling trees is considered unwise behaviour. With the small human numbers in the past, throwing away a little household degradable garbage could not have been considered wrong. But with enormous numbers of people throwing away large quantities of non-degradable waste, it is indeed extremely damaging to the environment and our value system must prevent this through a strong environmental value education system.

Appreciating the negative effects of our actions on the environment must become a part of our day to day thinking. Our current value system extols economic and technical progress as being what we need in our developing country.

## ENVIRONMENTAL VALUES BASED ON THE CONSTITUTION OF INDIA

- *Article 48A*: "The state shall endeavour to protect and improve the environment and to safeguard the forests and wildlife in the country."
- *Article 51A (g)*: The constitution expects that each citizen of the country must "protect and improve the natural environment, including forests, lakes, rivers and wildlife, and to have compassion for all living creatures."

While we do need economic development, our value system must change to one that makes people everywhere support a sustainable form of development so that we do not have to bear the cost of environmental degradation. Environmental problems created by development are due neither to the need for economic development, nor to the technology that produces pollution, but rather to a lack of awareness of the consequences of unlimited and unrestrained anti-environmental behaviour.

Looked at in this way, it deals with concepts of what is appropriate behaviour in relation to our surroundings and to other species on Earth. How we live our lives in fact shapes our environment. This is what environmental values are about. Each action by an individual must be linked to its environmental consequences in his/her mind so that a value is created that leads to strengthening pro-environmental behaviour and preventing anti-environmental actions.

This cannot happen unless new educational processes are created that provide a meaning to what is taught at school and college level. Every small child while growing up asks questions like 'What does this mean?'. They want an explanation for things happening around them that can help them make decisions and through this process develop values. It is this innate curiosity that leads to a personalized set of values in later life. Providing appropriate 'meanings' for such questions related to our own environment brings in a set of values that most people in society begin to accept as a norm. Thus pro environmental actions begin to move from the domain of individuals to that of a community. At the community level, this occurs only when a critical number of people become environmentally conscious so that they constitute a proenvironment lobby force that makes governments and other people accept good environmental behaviour as an important part of development.

What professions require making value judgements that greatly influence our environment? Evidentally nearly every profession can and does influence our environment, but some do so more than others. Policy makers, administrators, landuse planners, media, architects, medical personnel, health care workers, agriculturalists, agricultural experts, irrigation planners, mining experts, foresters, forest planners, industrialists and, most importantly, teachers at school and college level, are all closely related to pro environmental outcomes. Environmental values have linkages to varied enviro-nmental concerns. While we value resources that we use as food, water and other products, there are also environmental services that we must appreciate. These include Nature's mechanisms in cleaning up air by removing carbon dioxide and adding oxygen by plant life, recycling water through the water cycle of nature, maintaining climate regimes, etc. But there are other aesthetic, ethical values that are equally important aspects of our environment that we do not appreciate consciously.

While every species is of importance in the web of life, there are some which man has come to admire for their beauty alone. The tiger's magnificence, the whale and elephant's giant size, the intelligence of our cousins the primates, the graceful flight of a flock of cranes, are parts of nature that we cannot help but admire. The lush splendor of an evergreen forest, the great power of the ocean's waves, and the tranquility of the Himalayan mountains are things that each of us values even if we do not experience it ourselves. We value its being there on Earth for us. This is called its 'existance value'. The list of wondrous aspects of Nature's intricate connections is indeed awe-inspiring. This is also a part of our environment that we must value for its own sake. This is the oneness of Nature. We must equally look at our environment beyond the wild sphere. There is incredible beauty in some man-modified landscapes, the coloured patterns of farmland or the greens of a tea or coffee plantation in the hills.

Urban gardens and open space are also valuable and thus must be of prime concern to urban planners. These green spaces act as not only the lungs of a city, but also provide much needed psychological support. The mental peace and relaxation provided by such areas needs to be valued, although it is difficult to put a price tag on these values.

Nevertheless, these centers of peace and tranquility give urban dwellers an opportunity to balance their highly man-modified environments with the splash of green of a garden space. Environmental values must also stress on the importance of preserving ancient structures. The characteristic architecture, sculpture, artworks and crafts of ancient cultures is an invaluable environmental asset.

It tells us where we have come from, where we are now, and perhaps where we should go. Architectural heritage goes beyond preserving old buildings, to conserving whole traditional landscapes in rural areas and streetscapes in urban settings. Unless we learn to value these landscapes, they will disappear and our heritage will be lost.

As environmentally conscious individuals we need to develop a sense of values that are linked with a better and more sustainable way of life for all people. There are several positive as well as negative aspects of behaviour that are linked to our environment. The positive feelings that support environment include a value for Nature, cultures, heritage, and equity. We also need to become more sensitive to aspects that have negative impacts on the environment. These include our attitude towards degradation of the environment, loss of species, pollution, poverty, corruption in environmental management, the rights of future generations and animal rights. Several great philosophers have thoughts that have been based on, or embedded, in pro environmental behaviour.

Mahatma Gandhi and Rabindranath Tagore are among the many internationally well-known scholars whose thought have included values that are related to environmental consciousness. We need to appreciate these values to bring about a better way of life on earth for all people and all living creatures.

## VALUING NATURE

The most fundamental environmental sentiment is to value Nature herself. Appreciating Her magnificence and treasuring life itself leads to positive feelings that are a manifestation of pro environmental consciousness. The one-ness of our lives with the rest of nature and a feeling that we are only a miniscule part of nature's complex web of life becomes apparent, when we begin to appreciate the wonders of nature's diversity. We must appreciate that we belong to a global community that includes another 1.8 million known living forms.

Nothing makes us more conscious of this wonderous aspect of our earth's diversity than a walk through the wilderness, feeling and exploring its beauty and experiencing its infinite variety. The tiny creatures that live complex lives and the towering trees are all a part of this phenomenon we call 'life'. Today, man does not even know if other complex forms of life exist outside our own solar system in distant space. We may be alone in space or may be accompanied by other, completely different, living forms. But for now we only know for sure that the Earth's life forms are unique.

We thus have a great responsibility to protect life in all its glorious forms and must therefore respect the wilderness with all its living creatures, where man's own hand has not created changes that have led to perturbing natural habitats. We need to develop a sense of values that lead us to protect what is left of the wilderness by creating effective National Parks and Wildlife Sanctuaries.

However this cannot be done to the detriment of the millions of tribal or indigenous people who live in wilderness ecosystems. There are thus conflicting values that need to be balanced carefully. On the one hand we need to protect natural ecosystems, while on the other, we must protect the rights of local people. Yet apart from valuing the diversity of life itself, we must also learn to value and respect diverse human cultures. Many of the tribal cultures of our country are vanishing because those with more dominant and economically advanced ways of life do not respect their lifestyles, that are in fact closer to nature and frequently more sustainable. We believe that our modern technology-based lifestyles are the sole way for society to progress. Yet this is only a single dimension of life that is based on economic growth.

## DEEP ECOLOGY

In the 1970s a new thinking on environmental concerns began to emerge, protecting nature and the wilderness for its own sake, which is now referred to as 'Deep Ecology'. It was fostered by the thinking of Arne Naess, a Norwegian professor of Philosophy and a great believer in Gandhian thinking and Buddhism. It recognises the intrinsic value of all living beings and looks upon mankind as a small segment of a great living community of life forms.

It teaches that the wellbeing and flourishing of human and non-human life on Earth have value in themselves and that these values are independent of the usefulness of the non-human world for human purposes.

While currently the environmental movement focuses on issues that are concerned with the management of the natural environment for the 'benefit' of man, Deep Ecology promotes an approach that is expected to bring about a more appropriate ecological balance on Earth and is akin to a spiritual approach to Nature. This has great long-term implications not only for humans but for the whole of Nature. For example some environmentalists emphasise the need to preserve wilderness for its aesthetic and utilitarian functions.

Wilderness is being preserved today in PAs because it is scenic and serves the purpose of tourism for nature lovers, and has recreational and economic value. Other environmentalists stress that the goal is for protecting the useful ecological functions of the wilderness, its services and goods that we use. Deep Ecologists on the other hand stress that wilderness preservation is a means to achieve the conservation and protection of biological diversity. Thus it is not enough to protect bits of what is left of the wilderness but to make attempts to restore degraded areas to their former natural ecological state.

In a country such as India, with its enormous population coupled with poverty on the one hand and the need for economic industrial growth on the other, this will be extremely difficult to achieve. Another new approach is that of 'Gaia', the hypothesis that the Earth is itself like one giant form of throbbing life consisting of all the unquantifiable numbers of individuals of its millions of known and unknown species.

## VALUING CULTURES

Every culture has a right to exist. Tribal people are frequently most closely linked with Nature and we have no right to foist on them our own modern way of life. The dilemma is how to provide them with modern health care and education that gives them an opportunity to achieve a better economic status without disrupting their culture and way of life. This will happen only if we value their culture and respect their way of life.

## SOCIAL JUSTICE

As the divide widens between those people who have access to resources and wealth, and those who live near or below the poverty line, it is the duty of those who are better off to protect the rights of the poor who do not have the means to fight for their rights. If this is not respected the poor will eventually rebel, anarchy and terrorism will spread and the people who are impoverished will eventually form a desperate seething revolution to better their own lot.

The developing world would face a crisis earlier than the developed countries unless the rights of poor people that are fundamental to life are protected. Modern civilization is a homogenous culture, based until recently on a belief that modern science holds the answer to everything. We are now beginning to appreciate that many ancient and even present day sequestrated cultures have a wisdom and knowledge of their own environments that is based on a deep sense of respect for nature. Tribal cultures have over many generations used indigenous medicines which are proving to be effective against diseases. They have produced unique art forms such as painting, sculpture, and crafts that are beautiful and can enrich living experiences for everyone. They have their own poetry, songs, dance and drama-all art forms that are unfortunately being rapidly lost as we introduce a different set of modern values to them through television and other mass media.

The world will be culturally impoverished if we allow these indigenous people to loose their traditional knowledge which includes sustainable use of water, land and resources with a low impact on biodiversity. They will soon lose the beauty within their homes that is based on the things they make from Nature.

The art of the potter will be lost forever to the indestructible plastic pot. The bamboo basket weaver who makes a thing of beauty that is so user friendly and aesthetically appealing, will give place to yet another plastic box. Much that is beautiful and hand-crafted will disappear if we do not value these diverse aspects of human cultures.

## HUMAN HERITAGE

The earth itself is a heritage left to us by our ancestors for not only our own use but for the generations to come. There is much that is beautiful on our Earth-the undisturbed wilderness, a traditional rural landscape, the architecture of a traditional village or town, and the value of a historical monument or place of worship. These are all part of human heritage. Heritage preservation is now a growing environmental concern because much of this heritage has been undervalued during the last several decades and is vanishing at an astonishing pace. While we admire and value the Ajanta and Ellora Caves, the temples of the 10th to 15th centuries that led to different and diverse styles of architecture and sculpture, the Moghul styles that led to structures such as the Taj Mahal, or the unique environmentally-friendly Colonial buildings, we have done little to actively preserve them. As environmentally conscious individuals we need to lobby for the protection of the wilderness and our glorious architectural heritage.

## EQUITABLE USE OF RESOURCES

An unfair distribution of wealth and resources, based on a world that is essentially only for the rich, will bring about a disaster of unprecedented proportions. Equitable use of resources is now seen as an essential aspect of human well being and must become a shared point of view among all socially and environmentally conscious individuals.

This includes an appreciation of the fact that economically advanced countries and the rich in even poor nations consume resources at much greater levels than the much larger poorer sectors of humanity in the developing world. In spite of the great number of people in the more populous developing countries, the smaller number of people in developed countries use more resources and energy than those in the developing world.

This is equally true of the small number of rich people in poor countries whose per capita use of energy and resources, and the generation of waste based on the one time use of disposable products, leads to great pressures on the environment. The poor while polluting the environment have no way to prevent it.

The rich damage the environment through a carelessness that proves only that they have no appreciation for environmental safety. As we begin to appreciate that we need more sustainable lifestyles we also begin to realise that this cannot be brought about without a more equitable use of resources.

## COMMON PROPERTY RESOURCES

Our environment has a major component that does not belong to individuals. There are several commonly owned resources that all of us use as a community. The water that nature recycles, the air that we all breathe, the forests and grasslands which maintain our climate and soil, are all common property resources. When Government took over the control of community forests in British times, the local people who until then had controlled their use through a set of norms that were based on equitable use, began to overexploit resources on which they now had no personal stake.

Bringing back such traditional management systems is extremely difficult. However, in the recent past managing local forests through village level forest protection committees has shown that if people know that they can benefit from the forests, they will begin to protect them. This essentially means sharing the power to control forests between the Forest Department and local people.

## ECOLOGICAL DEGRADATION

In many situations valuable ecological assets are turned into serious environmental problems. This is because we as a society do not strongly resist forces that bring about ecological degradation. These consist of sectors of society that use a 'get-rich-quick' approach to development. While ecological degradation has frequently been blamed on the needs of fuelwood and fodder of growing numbers of rural people, the rich, urbanized, industrial sector is responsible for greater ecological damage.

Changes in landuse from natural ecosystems to more intensive utilization such as turning forests into monoculture forestry plantations, or tea and coffee estates, or marginal lands into intensive agricultural patterns such as sugarcane fields or changes into urban or industrial land carry an ecological price. Wetlands, for example, provide usable resources and a variety of services not easily valued in economic terms, and when destroyed to provide additional farmland, in many cases produce lower returns.

A natural forest provides valuable non-wood forest products whose economic returns far outweigh that provided by felling the forest for timber. These values must form a part of a new conservation ethic. We cannot permit unsustainable development to run onwards at a pace in which our lives will be overtaken by a development strategy that must eventually fail as Earth's resources are consumed and ecosystems rendered irreparable.

## HIV/AIDS

The Human Immunodeficiency Virus (HIV) causes Acquired Immunodeficiency Syndrome (AIDS) through contact with tissue fluids of infected individuals, especially through sexual contact. As it reduces an individual's resistance to disease, it causes infected individuals to suffer from a large number of environment related diseases and reduces the ability of infected individuals to go about their normal lives. It affects their income generation and/or their ability to utilise natural resources.

As more and more people are affected, this disease will also have impacts on our natural resource base, as utilisation patterns change to unsustainable levels. The inability of these patients to have the strength to access natural resources also affects the outcome of the disease process, as their overall health and well being is likely to worsen the course of the disease when their nutritional status suffers. In sub Saharan Africa where the infection has become highly prevalent, it is leading to great suffering and worsening poverty.

The capacity of these patients to work for their usual sources of income generation is lost. An increasing proportion of the poor are affected. It is evident that it is going to be increasingly difficult to manage environments sustainably, as natural resources on which the poor debilitated patients depend continue to be degraded. Incomes lost due to the stigma of HIV/AIDS must be met by the sufferers by overexploiting their resource base. People affected by the disease inevitably try to get whatever they can from their natural resource base as they are not in any position to think of the long-term future.

In Africa, this has led to degradation of the ecosystem and an increase of pressures from other impacts such as overuse of medicinal plants and poaching for wildlife. In South Africa, for example, people have a mistaken belief that turtle eggs can cure HIV/AIDS, thus leading to the eggs being over harvested. As males die of the disease, work on agricultural land has to be taken over by already overworked women and their children, affecting land management and productivity.

Providing balanced diets and nutritional support for these poverty stricken patients can be partially addressed by better natural resource management such as afforestation, access to clean water and wholesome food. HIV/AIDS seriously affects the patient's working environment. It creates an incorrect fear in the minds of co-workers. It must be clearly understood that AIDS is not spread by casual contact during work. Patients have a right to continue to work as before along with unaffected individuals. As patients are unable to continue their original hard labour related work, it is essential that alternative sources of work must be created for them. Educators and extention information, in the formal and non-formal educational sectors, must address the issues related to the linkages between natural resource management and this disease, as well as the need to remove the social stigma attached to it.

HIV/AIDS has a serious impact on the socioeconomic fabric of society. By 2002, India had an estimated 3.97 million infected individuals. There is a great need to organise AIDS education on prevention and management of the disease.

This needs to be done through the formal educational sector and by using non-formal methods. Education is also important to reduce the stigma and discrimination against these patients. In India, women who are not socially empowered are at a great disadvantage as they are powerless to demand safe sex from their partners.

Women also have an added burden of caring for HIV infected husbands. This produces enormous economic stresses on their family. HIV in India is rapidly moving from a primarily urban sector disease to rural communities. Research in Nepal has shown a linkage between rural poverty, deforestation and a shift of population to urban areas resulting in a rising number of AIDS patients.

Prior to 1992, it was mainly seen in males who migrated to urban centers. In more recent times, a growing number of women are moving to Indian cities as sex workers. Women engaged in prostitution find it difficult to make partners take protective measures, such as the use of condoms that provide safe sex. A large proportion became victims of the disease. Blood transfusion from an infected person can also lead to HIV/AIDS in the recipient, as well as drug abuse by sharing needles with an infected person.

In sexually transmitted AIDS, the use of condoms during intercourse is a key to preventing the disease. Behavioural change, where the number of individuals who have multiple partners, towards strictly single partners, reduces the risk of HIV/AIDS and thus reduces incidence of the disease in society. However, the most important measure to prevent AIDS is the proper use of condoms that form a barrier to the spread of the virus during intercourse.

## WOMAN AND CHILD WELFARE

There are several environmental factors that are closely linked to the welfare of women and children. Each year, close to eleven million children worldwide are estimated to have died from the effects of disease and inadequate nutrition. Most of these deaths are in the developing world. In some countries, more than one in five children die before they are 5 years old. Seven out of 10 of childhood deaths in developing countries can be attributed to five main causes, or a combination of them. These are pneumonia, diarrhoea, measles, malaria and malnutrition. Around the world, three out of every four children suffer from at least one of these conditions.

**Table. The Diagnosis of Common Childhood Disease Conditions**

| Presenting complaint | Possible cause or associated condition |
|---|---|
| Cough and/or fast breathing | Pneumonia<br>Severe anaemia<br>P. falciparum malaria |
| Lethargy or unconsciousness | Cerebral malaria<br>Meningitis<br>Severe dehydration<br>Very severe pneumonia |
| Measles rash | Pneumonia<br>Diarrhoea<br>Ear infection |
| "Very sick" young infant | Pneumonia<br>Meningitis<br>Sepsis |

## RESPIRATORY CONDITIONS

Most respiratory diseases are caused by or are worsened by polluted air. Crowded ill-ventilated homes and living in smokey households with open fires can trigger respiratory conditions especially in children. Pneumonia: Acute respiratory infections (ARI), most frequently pneumonia, is a major cause of death in children under five, killing over two million children annually. Upto 40 per cent of children seen in health centers suffer from respiratory conditions and many deaths attributed to other causes are, in fact, "hidden" ARI deaths. Children may die very quickly from the infection and thus need treatment urgently. Most patients of pneumonia can be treated with oral antibiotics. Correct management could save over 1 million lives per year globally.

## GASTRO INTESTINAL CONDITIONS

Contaminated water and food causes widespread ill health especially in children. Diarrhoea: Diarrhoea is caused by a wide variety of infections. Urgent diagnosis and treatment of diarrhoea is a priority for saving a child's life. Treating malnutrition that often accompanies diarrhoea can further reduce mortality. Increasing vigilance to detect other diseases that can occur concurrently with diarrhoea, such as measles or malaria, is an important measure.

Two million children die each year in developing countries from diarrhoeal diseases, the second most serious killer of children under five worldwide. In most cases diarrhoea is preventable and children can be saved by early treatment. Correct management of diarrhoea could save the lives of up to 90 per cent of children who currently die by promoting rapid and effective treatment through standardised management, including antibiotics and simple measures such as oral rehydration using clean boiled water with salt and sugar.

In severe cases intravenous fluids must be started. Improved hygiene and management of the home and surroundings is the most important preventative measure, as well as improved nutrition. Increased breastfeeding and measles vaccination have also been observed to have reduced the number of cases of diarrhoea.

## MEASLES

Measles is a rash that appears with fever and bodyache in children and is caused by a virus. It infects over 40 million children and kills over 800,000 children under the age of five. Prevention includes wider immunization coverage, rapid referral of serious cases, prompt recognition of conditions that occur in association with measles, and improved nutrition, including breastfeeding, and vitamin A supplementation. Measles is prevented by a vaccine. Young children with measles often develop other diseases such as acute respiratory infections, diarrhoea and malnutrition that are all linked to poor environmental conditions in their surroundings. Children who survive an attack of measles are more vulnerable to other dangerous infections for several months. Effective prevention and treatment could save 700,000 lives per year.

## MALARIA

This condition is closely linked to pooling and stagnation of water in tropical environments. Malaria is a widespread tropical disease which is caused by a parasite transmitted to humans by mosquitoes. It has proved difficult to control because mosquitoes have become resistant to insecticides used against them and because the parasite has developed resistance in some areas to the cheap and effective drugs that used to provide good protection in the past.

However, alternative newer drug therapies have been developed for use in areas where resistant parasites are found. In India the disease was nearly wiped out a few decades ago but has now re-emerged in many parts of the country. Correct management could save 500,000 lives per year. Approximately 700,000 children die of malaria globally each year, most of them in sub-Saharan Africa.

Young children are particularly vulnerable because they have not developed the partial immunity that results from surviving repeated infections. Deaths from malaria can be reduced by several measures, including encouraging parents to seek prompt care, accurate assessment of the condition of the child, prompt treatment with appropriate anti-malarial drugs, recognition and treatment of other co-existing conditions, such as malnutrition and anaemia, and prevention by using mosquito-proof bednets. Because fever may be the only sign of malaria, it can be difficult to distinguish it from other potentially lifethreatening conditions.

## POVERTY-ENVIRONMENT-MALNUTRITION

There is a close association between poverty, a degraded environment, and malnutrition. This is further aggravated by a lack of awareness on how children become malnourished. Malnutrition: Although malnutrition is rarely listed as the direct cause of death, it contributes to about half of all childhood deaths. Lack of access to food, poor feeding practices and infection, or a combination of the two, are major factors in mortality.

Infection, particularly frequent or persistent diarrhoea, pneumonia, measles and malaria, undermines nutritional status. Poor feeding practices-inadequate breastfeeding, providing the wrong foods, giving food in insufficient quantities, contribute to malnutrition. Malnourished children are more vulnerable to disease. Promoting breastfeeding, improving feeding practices, and providing micronutrient supplements routinely for children who need them are measures that reduce mortality.

The nutritional status and feeding practices of every child under two years of age, and those with a low weight for their age must be intensively managed. Counseling of parents on the correct foods for each age group and helping them to overcome various feeding problems is an essential health care measure. Children between 6 months and 2 years of age are at increased risk of malnutrition when there is a transition between breastfeeding and sharing fully in the family diet.

Changing family habits and the kinds of food offered to children is an important measure. Talking to mothers individually about home care and their child's feeding, with relatively simple changes to better feeding practices, such as helping them to eat rather than leaving them to fend for themselves, can ensure that a child gets enough to eat. A minor increase in breastfeeding could prevent up to 10 per cent of all deaths of children under five: When mothers breastfeed exclusively during at least the first four months and, if possible, six months of life, there is a decrease in episodes of diarrhoea and, to a lesser extent, respiratory infections.

Even small amounts of water-based drinks decreases breastmilk intake and lead to lowered weight gain. This increases the risk of diarrhoea. Continuing to breastfeed up to two years of age, in addition to giving complementary foods, maintains good nutritional status and helps prevent diarrhoea. Encouraging maximum support to mothers to establish optimal breastfeeding from birth, equipping health workers with counseling skills, and providing individual counseling and support for breastfeeding mothers are measures that reduce malnutrition.

Mothers often give their babies other food and fluids before six months because they doubt their breastmilk supply is adequate. A one-on-one counseling with mothers on breastfeeding techniques and its benefits helps reduce incidence of malnutrition. There are strong connections between the status of the environment and the welfare of women and children in India. Women, especially in lower income group families, both in the rural and urban sector, work longer hours than men.

Their work pattern differs and is more prone to health hazards. The daily collection of water, fuelwood and fodder is an arduous task for rural women. In urban areas, where lower economic group women live in crowded smoke filled shantys in unhygenic slums, they spend long hours indoors, which is a cause of respiratory diseases. In urban centers, a number of women eke out a living by garbage picking. They separate plastics, metal and other recyclable material from the waste produced by the more affluent groups of society. During this process, they can get several infections. Thus they are providing an environmental service of great value, but earn a pittance from this work.

Women are often the last to get enough nutrition as their role in traditional society is to cook the family meal and feed their husband and children. This leads to malnutrition and anemia due to inadequate nutrition. The sorry plight of women includes the fact that the girl child is given less attention and educational facilities as compared to boys in India. Thus

**CASE STUDY**

Karnataka's GIS scheme, Bhoomi, has revolutionized the way farmers access their land records. Farmers can now get a copy of the records of rights, tenancy and crops from a computerized information kiosk without harassment and bribes. Karnataka has computerized 20 million records of land ownership of 6.7 million farmers in the State. they are unable to compete with men in later life. This social-environmental divide is a major concern that needs to be corrected throughout the country.

## ROLE OF INFORMATION TECHNOLOGY IN ENVIRONMENT AND HUMAN HEALTH

The understanding of environmental concerns and issues related to human health has exploded during the last few years due to the sudden growth of Information Technology. The computer age has turned the world around due to the incredible rapidity with which IT spreads knowledge. IT can do several tasks extremely rapidly, accurately and spread the information through the world's networks of millions of computer systems.

A few examples of the use of computer technology that aid environmental studies include software such as using Geographical Information Systems (GIS). GIS is a tool to map landuse patterns and document change by studying digitized toposheets and/or satellite imagery. Once this is done, an expert can ask a variety of questions which the software can answer by producing maps which helps in landuse planning.

The Internet with its thousands of web sites has made it extremely simple to get the appropriate environmental information for any study or environmental management planning. This not only assists scientists and students but is a powerful tool to help increase public awareness about environmental issues. Specialised software can analyse data for epidemiological studies, population dynamics and a variety of key environmental concerns. The relationship between the environment and health has been established due to the growing utilisation of computer technology. This looks at infection rates, morbidity or mortality and the etiology (causative factors) of a disease. As knowledge expands, computers will become increasingly efficient. They will be faster, have greater memories and even perhaps begin to think for themselves.

# Bibliography


Anderson, J.T.: *Farm Forestry Practices for the Students of Vocational Agriculture*, Asiatic Pub, Delhi, 2008.

Anthony B.: *Applied Ethnobotany: People, wild Plant use, and Conservation*. London, Earthscan, 2001.

Babweteera, F.: *Influence of Gap-size and Age on the Diversity and Abundance of Climbers in Budongo Forest Reserve*, Uganda. M. Sc., Makerere University, 1998.

Bagyaraj, D. Joseph: *Microbial Biotechnology for Sustainable Agriculture Horticulture and Forestry*, New India Publishing Agency, Delhi, 2011.

Bahati, J.: *Impact of Arboricidal Treatments on the Natural Regeneration of Species Composition in Budongo Forest*, Uganda. M. Sc., Makerere University, 1995.

Bawa, R. and P.K. Khosla : *Biodiversity of Forest Types A Community Forestry Approach* , BSMPS, Delhi, 1998.

Bhattacharya, Ajoy Kumar: *Community Participation and Sustainable Forest Development*, Concept, Delhi, 2001.

Bisht, Anand Singh; Krishan Datt Sharma and Matthew Prasad: *Food and Horti-Forestry Resources of Western Himalaya*, Jaya, Delhi, 2014.

Bor, N. L.: *A Manual of Indian Forest Botany*, Asiatic Pub, Delhi, 2010.

Chakrabarty, Falguni: *Adaptation of the Santals to the Hill-Forest Environment*, APH, Delhi, 2012.

Connor Ogorzaly: *Economic Botany: Plants in our World*. Boston, McGraw-Hill, 2001.

Falguni, C: *Adaptation of the Santals to the Hill-Forest Environment*, APH, Delhi, 2012.

Jackie C. Wooten : *Ethnomedicine and Drug Discovery*, Amsterdam, New York, Elsevier, 2002.

Macself, A.J. : *Soils and Fertilizers*, Satish Serial Pub, Delhi, 2005.

Miller, P. L.: *Some Dragonflies of the Budongo Forest*, Western Uganda. Opusc, 1993.

Mohd. Azharul Haque and Tanmoy Rudra : *Biodiversity Extinction and Deforestation* , Jnanada Prakashan, Delhi, 2010.

Reynolds, V.: *Budongo: A Forest and its Chimpanzees*, London, Methuen, 1965.

Sathe, T.V.: *A Textbook of Forest Entomology*, Daya, Delhi, 2009.

Sett, Rupnarayan: *Environmental Science : Botanical and Forestry Perspective*, Narendra Pub, Delhi, 2012.

Sharma, Ajay: *A Text Book of Forest Trees Entomology*, International Book Dist, Delhi, 2007.

Sharma, R.C. and J.N. Sharma : *Challenging Problems in Horticultural and Forest Pathology*, Indus, Delhi, 2005.

Sheelwant Patel: *Indian Forests : Soil Water and Bio-Environment Conservation*, Pointer, Delhi, 2005.

Shiel, D.: *The Ecology of Long-term Change in a Ugandan Rainforest*, D. Phil., Oxford University, 1997.

Shiel, D.: *The Ecology of Long-term Change in a Ugandan Rainforest*. D. Phil., Oxford University, 1997.

Trivedi, P.C. : *Plant Physiology in Agriculture and Forestry*, Aavishkar Pub, Delhi, 2009.

Ward, H. Marshall: *Trees a Handbook of Forest-Botany*, Asiatic, Delhi, 2001.

Wickens, G. E.: *Economic Botany: Principles and Practices*. Dordrecht, Boston, Kluwer Academic, 2001.

# Index